Peter S. Hechl
Reuben C. Setliff, III
Manfred Tschabitscher

Endoscopic Anatomy of the Paranasal Sinuses

Springer-Verlag Wien GmbH

Dr. Peter S. Hechl
Krems, Austria

Dr. Reuben C. Setliff, III
North Platte, NE, U.S.A.

Prof. Dr. Manfred Tschabitscher
Institute of Anatomy, University of Vienna,
Vienna, Austria

Originally published by Springer-Verlag Wien New York in 1997

Typesetting and Artwork Reproduction:
Seyss GmbH • Pre-Press- & Medien-Services/Wien

Graphic design: Ecke Bonk

Printed on acid-free and chlorine-free bleached paper

Cover photo: Zero degree lens, left side, uncinate process partially removed, ethmoid bulla opened, and basal lamella visible (vertical and horizontal portions)

With 194 partly coloured Figures

Die Deutsche Bibliothek - CIP-Einheitsaufnahme
Hechl, Peter S.:
Endoscopic anatomy of the paranasal sinuses / Peter S. Hechl; Reuben C. Setliff, III; Manfred Tschabitscher. - Wien ; New York ; Springer, 1997
ISBN 978-3-211-82922-6 ISBN 978-3-7091-6536-2 (eBook)
DOI 10.1007/978-3-7091-6536-2

CIP data applied for

Forewords

The Department of Anatomy in Vienna is proud of a long tradition in clinical-orientated anatomy, which has been performed by a number of well-known anatomists. One can remember Zuckerkandl, who produced important books about surgical anatomy, and Tandler, who made valuable contributions to practical anatomy in different fields. Therefore, I am happy to add a few lines to a new project of applied anatomy, namely the atlas of endoscopic surgical anatomy of the paranasal sinuses by P. Hechl, R. C. Setliff, and M. Tschabitscher. The authors undertook painstaking preparations of the anatomical material. The development of new surgical techniques makes it necessary to look for the morphological background of such a difficult region. This volume is an important step towards the desired synthesis of modern surgical development and morphological investigations. I am convinced that this goal has been achieved to the advantage of modern medicine.

W. Firbas, Dr. Univ.-Prof.
Department Head
Institute of Anatomy
University of Vienna
Austria

*

The otolaryngologist thrives in the study of head and neck anatomy. Nowhere is this more true than in the fascinating anatomy of the paranasal sinuses as it relates to endoscopic sinus surgery. As surgeons we seek out from a wide variety of sources: a photograph or line drawing which furthers our understanding on this complex area.

It is fitting that this current endeavor to promote understanding of the anatomy of the ethmoid bone should come out of the Department of Anatomy in Vienna with its long and proud tradition of clinically oriented anatomy dating back to Zuckerkandl. Drs. Hechl and Setliff have successfully undertaken the difficult task of melding classic surgical anatomy with the needs of the endoscopic sinus surgeon. This volume is an important step forward in our understanding of the anatomy in a clinically useful form. We are convinced that the authors have exceeded their goal in this regard and that we as readers are all the better for it.

Mark May, M.D., F.A.C.S.
Barry Schaitkin, M.D.

*

The techniques of near bloodless surgery, minimal invasion of tissues, protecting the final common pathways, are advanced by Reuben Setliff. The visibility allowed by such precision surgery gives new and greater appreciation of anatomic subtleties, seldom appreciated by us commoners. This book is a collection of months of intense work to show otolaryngologists the subtleties in the paranasal sinuses so that we, too, can improve our skills.

All otolaryngologists share the same goal - to provide excellent surgical care to the suffering patients. The authors' dissections and photography provide a clearer understanding to the readers desiring to improve their knowledge of complex sinus anatomy.

David S. Parsons, M.D., F.A.A.P., F.A.C.S.
Professor of Surgery and Pediatrics
University of Missouri School of Medicine
Columbia, Missouri, USA

Preface

One of us, Peter Hechl, former department head (Department of Otorhinolaryngology, Krankenhaus Krems, Austria), conceived the notion of an endoscopic anatomy atlas for paranasal sinuses. Years of laborious investigation and dissection in cooperation with our anatomist, Prof. Manfred Tschabitscher, followed. Springer-Verlag Wien graciously supported the project. The motivation to bridge the space between the "plane of the anatomist" and the landscape confronting the endoscopist has been a driving force in the completion of the work.

Due to Hechl's interest in the work of the third collaborator, Reuben Setliff, an invitation to join the project was extended. Concepts of a systematic surgical approach using minimal invasive techniques, transition spaces, and a study of secretory pathways brought a wealth of surgical anatomy to complement the already available intensive laboratory and surgical experience.

For the beginner or for the accomplished sinus surgeon, mastering the anatomy of the lateral nasal wall is an ongoing challenge. Even though there are excellent standard anatomical references and equally outstanding sinus courses with cadaver dissection, a reference depicting the surgical anatomy is needed. This book is a product of the advent of endoscopic visualization and precision instruments, making a controlled anatomical dissection possible for the first time. A step-by-step surgical approach on the anterior nasal spine to the anterior wall of the sphenoid is presented.

The sinus surgeon is confronted with a wide range of different spaces created by the ethmoid bone. No other bone in the human body has so many anatomical variations. Four critical anatomical structures are emphasized as the foundation for a precise approach to surgery of the maxillary, anterior ethmoid, and frontal and posterior ethmoid sinuses.

The goal of this book is to meet the tremendous challenge of offering an anatomical approach which will serve the sinus surgeon of every level of experience and expertise. Acknowledging that much of our knowledge comes from contributions of others, it is our hope that this text will complement their work.

Peter S. Hechl
Reuben C. Setliff, III
Manfred Tschabitscher

Acknowledgments

Thanks so much to Raimund Petri-Wieder, Editorial Department, Springer-Verlag Wien, for accepting this project and encouraging us to finish this book.

A special thanks to Marge Hehnke, Gail Setliff, Shelly Young, Cheryl Jensen, and other members of Dr. Setliff's office staff for their time and effort, which exceeded reasonable expectations.

We want to thank Andreas Maurer, M.D., for his expertise in photography and his unfailing generosity in sharing his knowledge and time.

Thanks are also due to M. Baumann, Vienna, for her technical skills in preparing black and white pictures and for the many hours she spent on this project.

We are indebted to the expertise of Photolab Kadmon, Vienna, for preparation of excellent slides.

We also want to thank Professional Photo Laboratories under the direction of Jan Morgan in Marina del Rey, California.

Thanks to Elizabeth Prilisauer for helping us in numerous situations and typing.

Our appreciation also to Deanne Kloepfer for her dedication in making the book possible.

Thanks to Mr. Rudoletzky, Olympus, Vienna, for supplying an expensive photo system for photo documentation.

We are especially indebted to Ethicon, Vienna, for generous grants and contributions over a number of years that have made this work possible.

Personal acknowledgments

A special acknowledgment for the inspiration to enter the medical field and the encouragement given me by my parents, Drs. Lydia and Hans Hechl.

P.S.H.

Contributions in medicine can only be made in a background of knowledge and assistance gained from others. My mentor as a resident was Dr. Paul Ward, a unique combination of clinician and researcher who instilled in me an interest in teaching and new knowledge. Having learned from every endoscopic surgeon whom I have encountered, special thanks must go to Drs. David Parsons, Rod Lusk, and Mark May, all of whom have given me encouragement and invaluable background information.

R.C.S.

Contents

List of abbreviations

ACC	Accessory Ostium
AE	Anterior Ethmoid
AEA	Anterior Ethmoid Artery
AEO	Anterior Ethmoid Ostium
AN	Agger Nasi
ANC	Agger Nasi Cell
BL	Basal Lamella
C	Carotid
CB	Concha Bullosa
EB	Ethmoidal Bulla
FCP	Final Common Pathway
FS	Frontal Sinus
H	Haller Cell
HS	Hiatus Semilunaris
HSP	Hiatus Semilunaris Posterior
IM	Inferior Meatus
INF	Infundibulum
IT	Inferior Turbinate
LNW	Lateral Nasal Wall
LP	Lamina Papyracea
MM	Middle Meatus
MMA	Middle Meatal Antrostomy
MO	Maxillary Ostium
MS	Maxillary Sinus
MSO	Maxillary Sinus Ostium
MT	Middle Turbinate
NAW	Naso-Antral Window
NC	Nasal Cavity
NS	Nasal Septum
OFT	Outflow Tract
OG	Olfactory Groove
ON	Optic Nerve
P	Polyp
PE	Posterior Ethmoid
PF	Posterior Fontanelle
SB (RV)	Skull Base (Ribbed Vault)
SE	Septum
SER	Sphenoethmoid Recess
SL	Sinus Lateralis
SM	Superior Meatus
SS	Sphenoid Sinus
ST	Superior Turbinate
TU	Tumor
UP	Uncinate Process
V	Nasal Vestibule
X	Exit of Infundibulum

1. The nasal septum

The nasal septum is the partition that divides the nose into right and left nasal cavities. It is a composite of bony and cartilaginous contributions and is usually deviated out of the midline to some degree. Remarkably, a significant clinical deviation may be present without associated nasal symptoms.

For the sinus surgeon the deviation becomes quite significant, whether symptomatic or not, when it limits access to the nasal cavity and/or middle meatus. The deviation may be localized and amenable to a limited endoscopic procedure. A more extensive deviation may require a formal surgical approach.

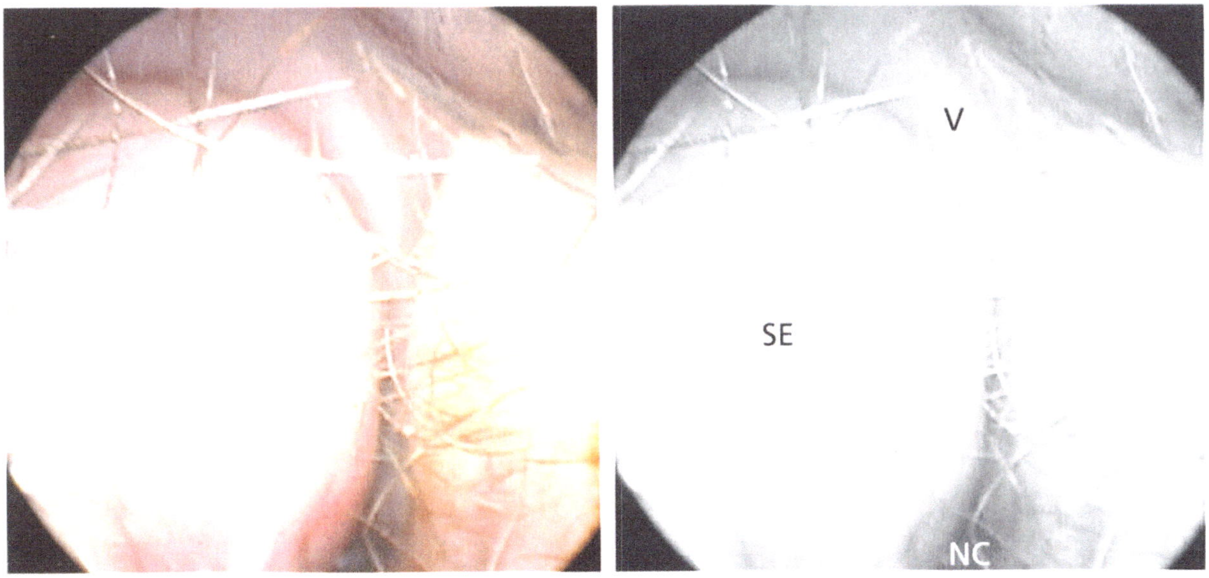

Fig. 1.1. Left side, zero degree lens, septal obstruction at nasal vestibule.

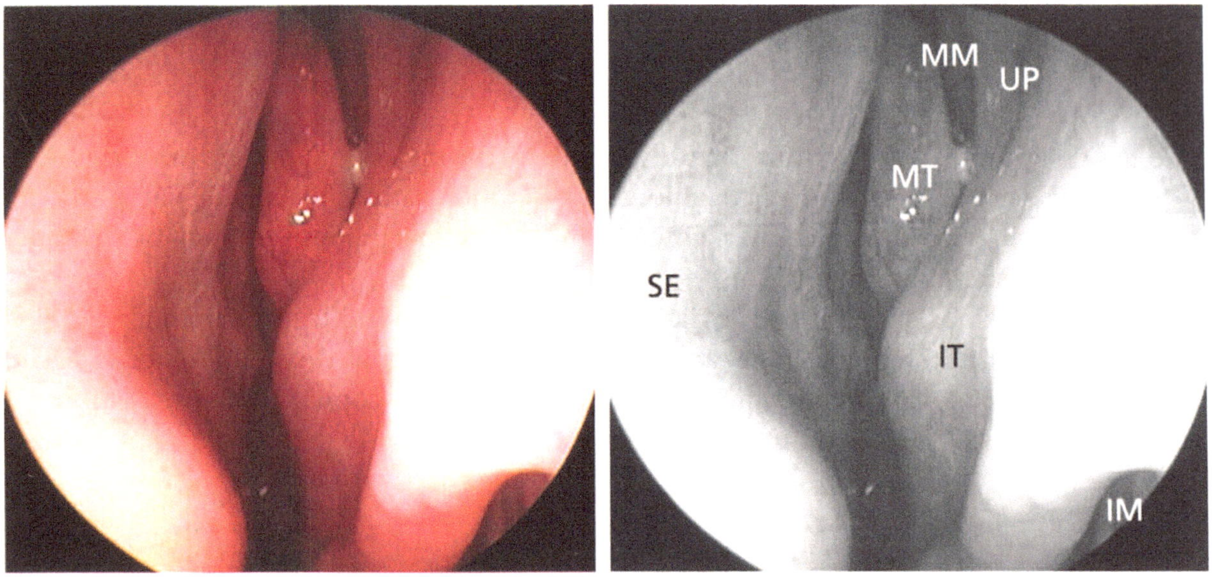

Fig. 1.2. Left side, zero degree lens, no significant septal deviation.

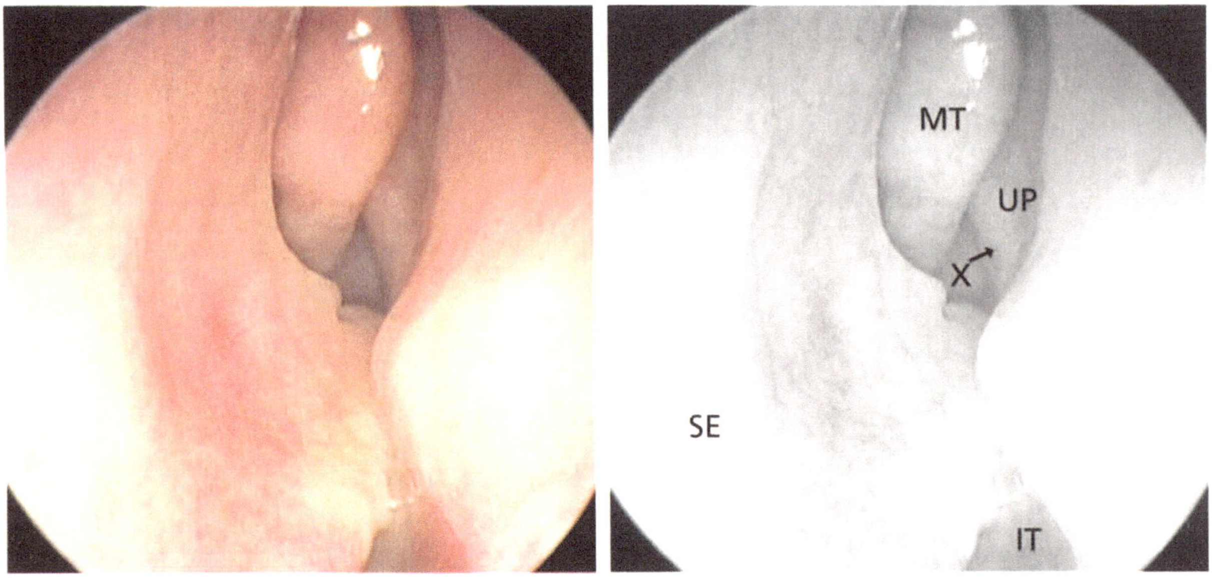

Fig. 1.3. Left side, zero degree lens, septal deviation with spur pushing into lateral nasal wall. Middle turbinate above.

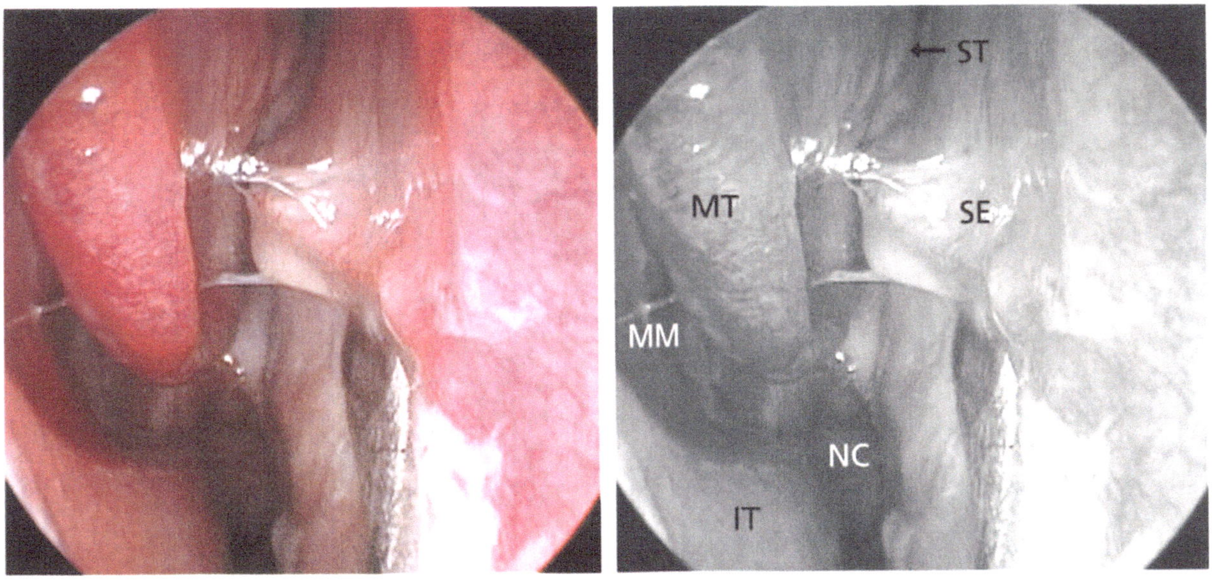

Fig. 1.4. Right side, zero degree lens, indentation in middle turbinate from septal spur; Exposed by Adson elevator.

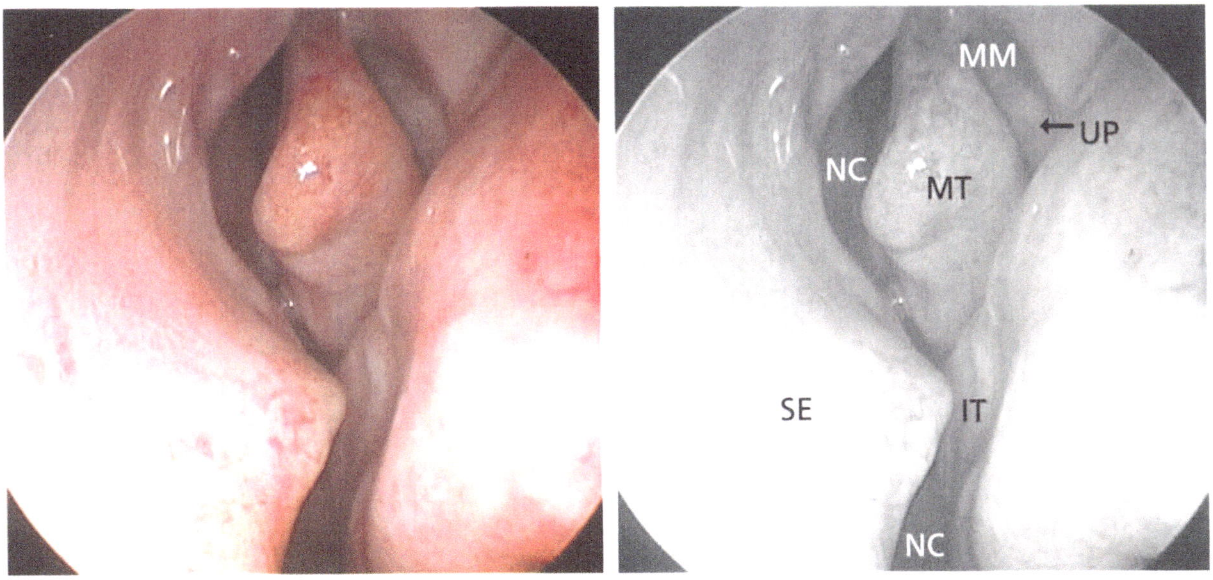

Fig. 1.5. Left side, zero degree lens, indentation in inferior turbinate from septal spur, paradoxical and pneumatized middle turbinate above.

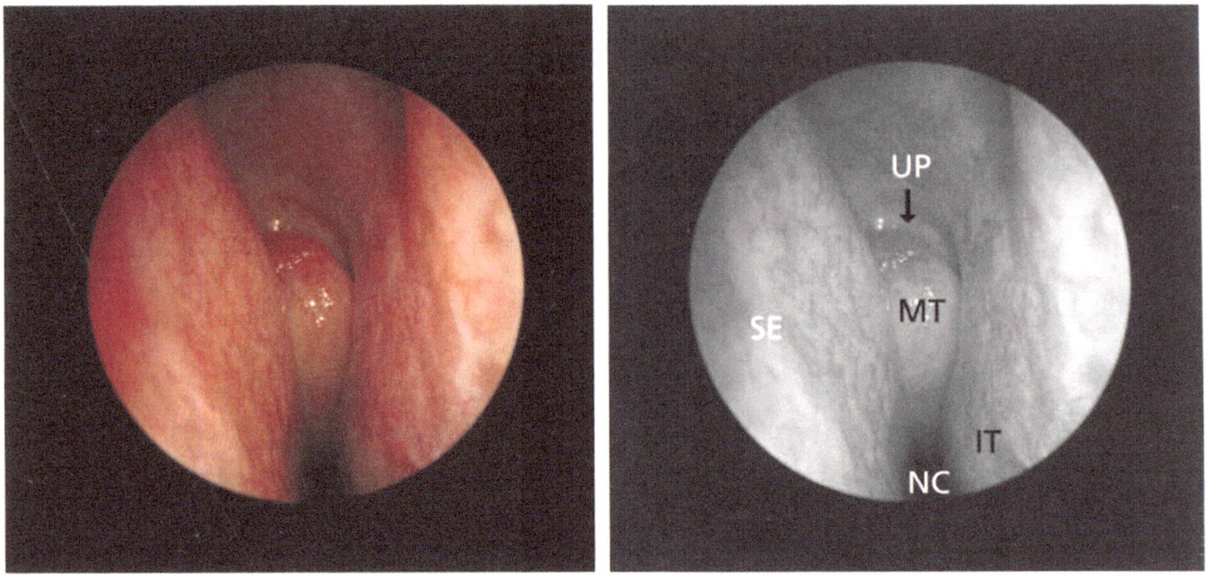

Fig. 1.6. Left side, zero degree lens, bulbous deviation pushing turbinate laterally, middle meatus partially closed.

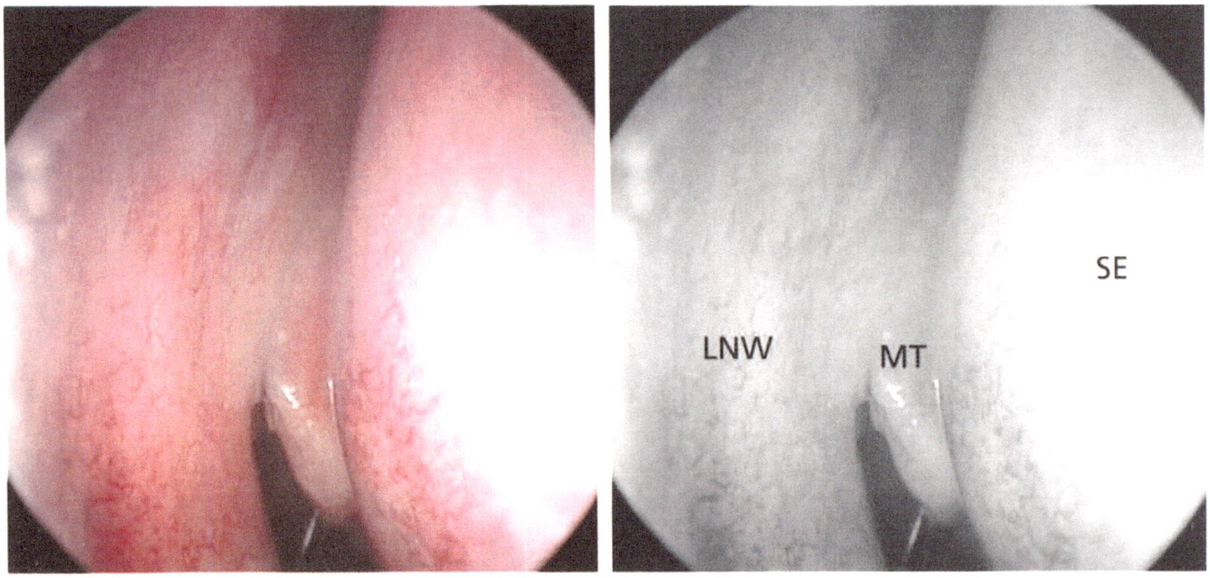

Fig. 1.7. Right side, zero degree lens, middle turbinate squeezed between nasal septum and lateral nasal wall, partial closure of middle meatus.

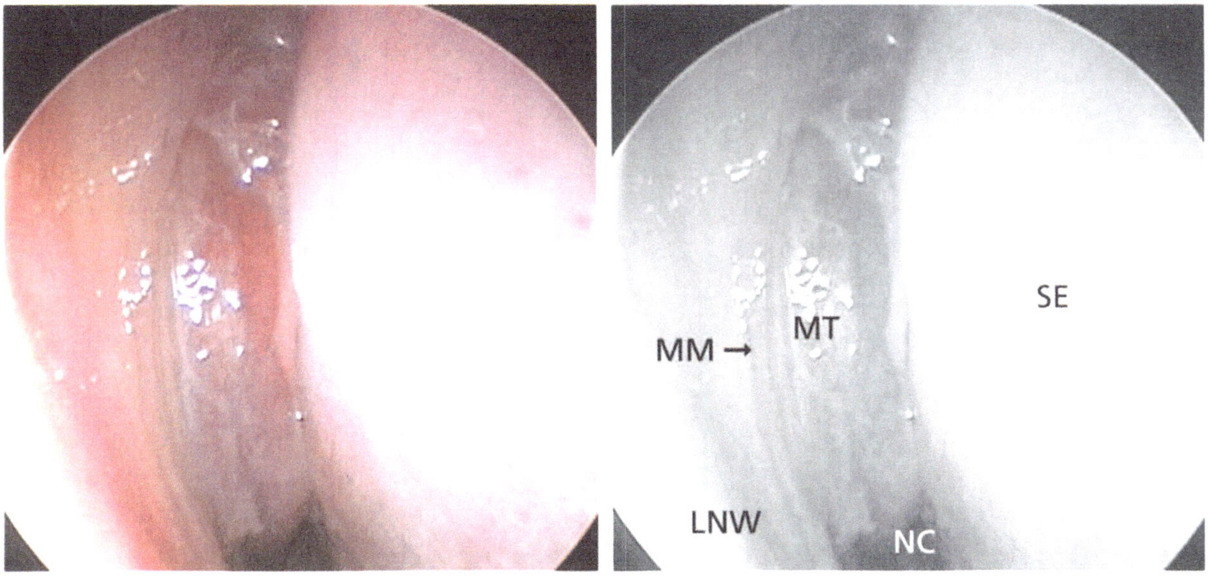

Fig. 1.8. Right side, zero degree lens, marked deviation with middle turbinate contact and polyp with middle meatus closed.

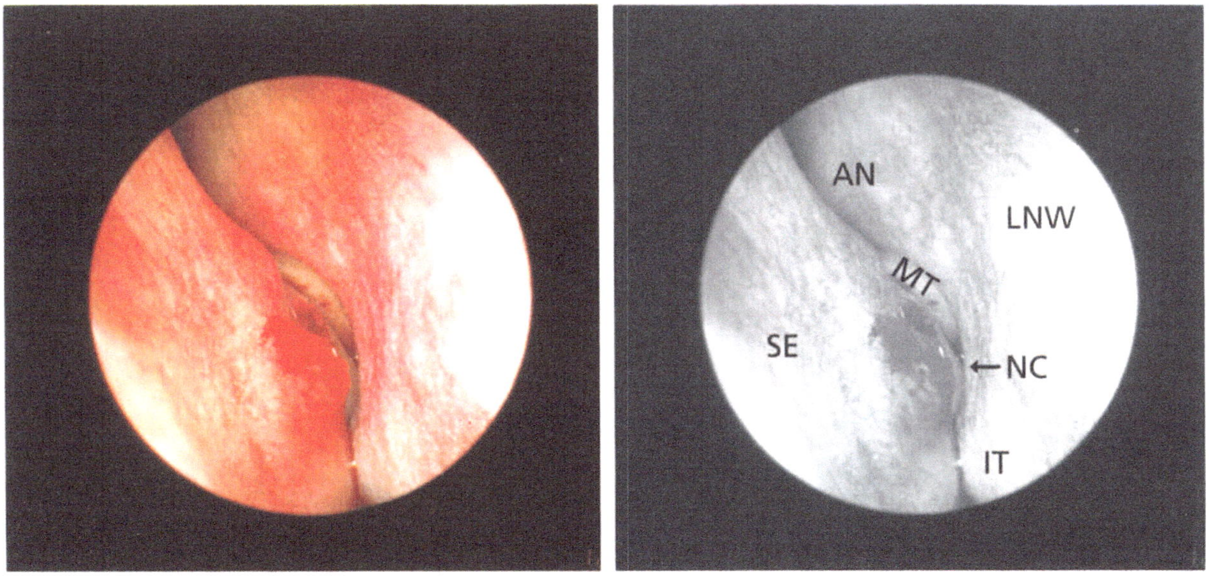

Fig. 1.9. Left side, zero degree lens, obstructing convexity.

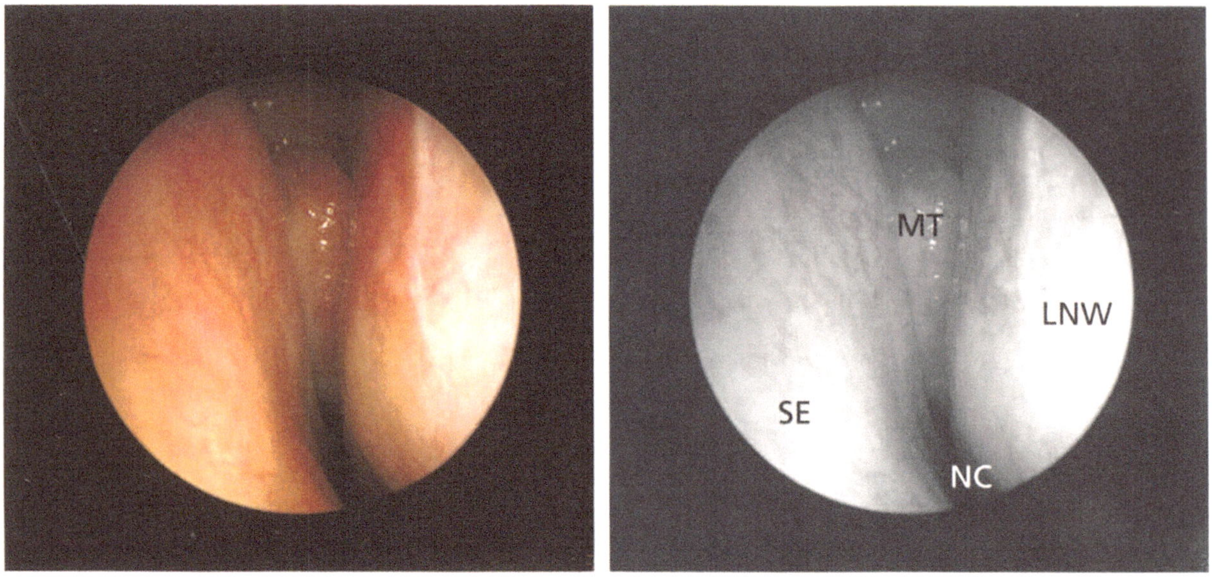

Fig. 1.10. Left side, zero degree lens, obstructing convexity of nasal septum, middle turbinate and uncinate process barely visible.

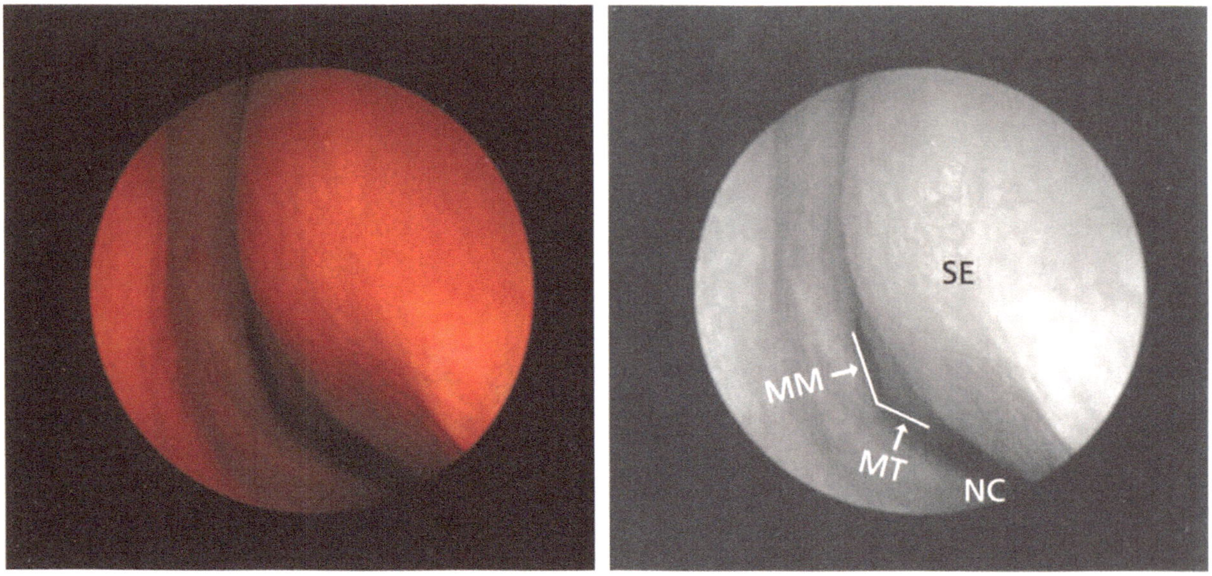

Fig. 1.11. Right side, zero degree lens, middle turbinate squeezed between septum and lateral nasal wall.

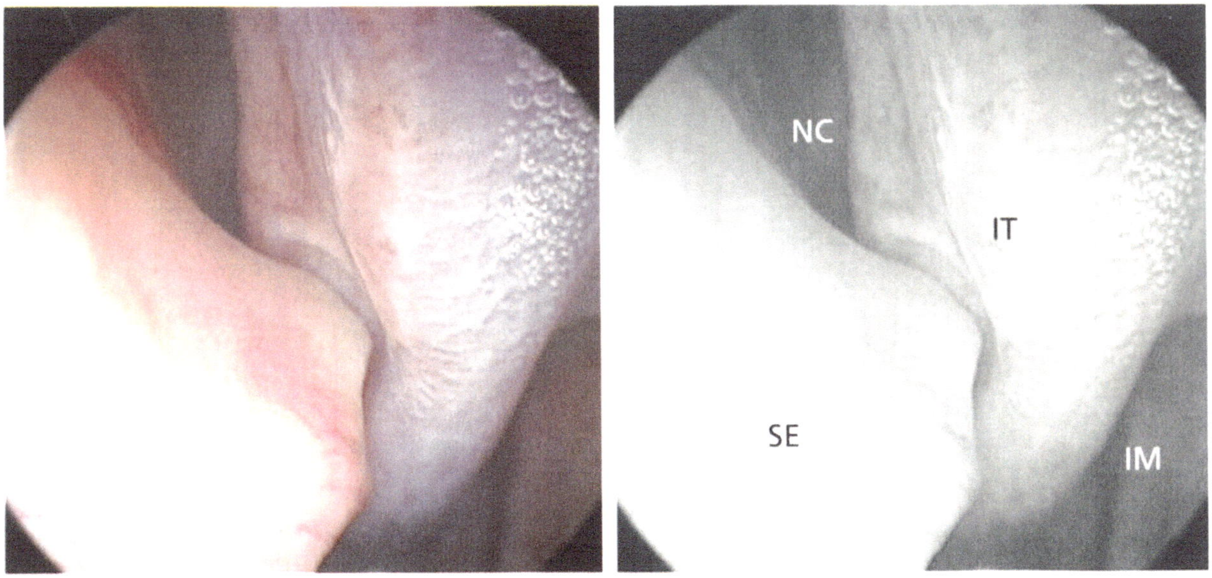

Fig. 1.12. Left side, zero degree lens, septal spur indenting inferior turbinate.

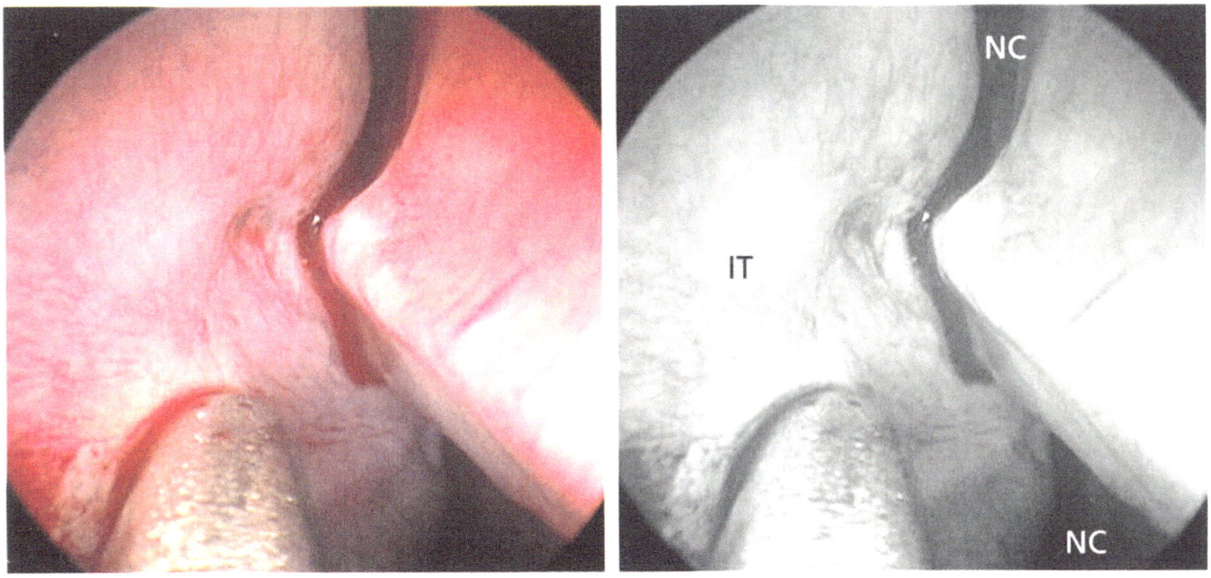

Fig. 1.13. Right side, zero degree lens, same as Fig. 1.12.

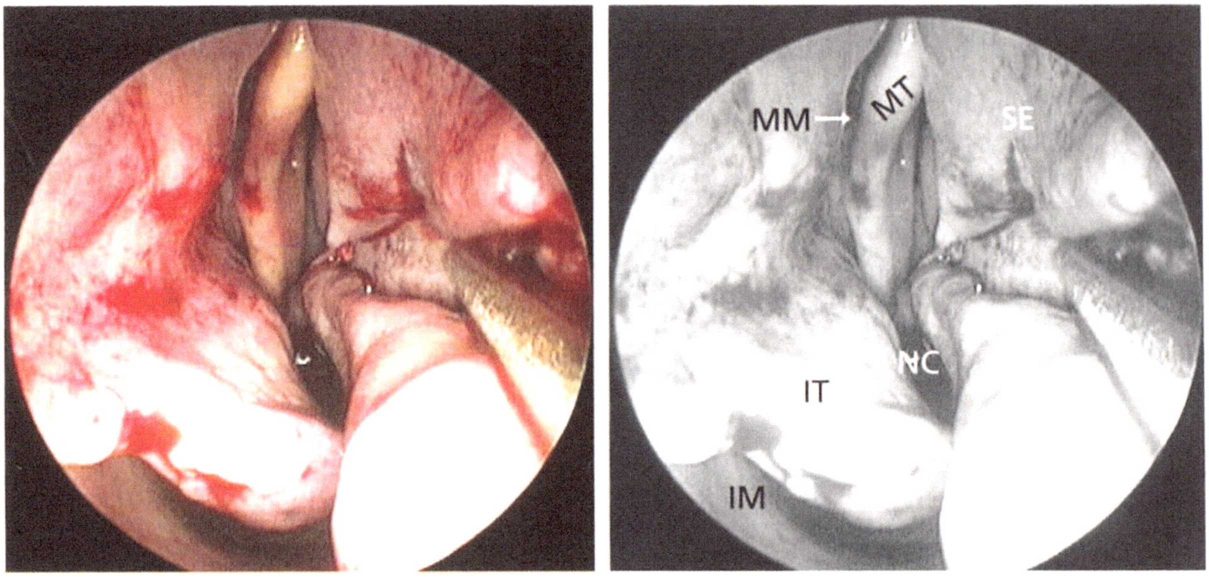

Fig. 1.14. Right side, zero degree lens, marked indentation of the inferior turbinate as seen following endoscopic resection of septal deviation and spur.

2. The ethmoid bone and middle turbinate

The ethmoid bone is a highly differentiated bone of the cranium. It is part of the anterior skull base, nasal skeleton, medial wall of the orbit, the medial wall of the maxillary sinus and must be considered the most important part of the sinus system.

It is filled with very differently shaped, partially opened cells (spaces) and with a complicated system of partitions (grand lamellae). All cells and spaces have openings or ostia. Although both the anatomist and the endoscopist refer to individual components of the ethmoid bone as separate structures, there is an ultimate continuity of all parts. The ethmoid component that is most difficult to understand is the basal lamella of the otolaryngologist, also known as the third grand lamella of the anatomist.

Most of the basal lamella is inside the cell system of the ethmoid labyrinth. It is continuous with both anterior and posterior ethmoid cells and is referenced to mark the division between the cells. It is also continuous with the middle turbinate medially, becoming the bony part of the middle turbinate itself. Laterally it is continuous with the lamina papyracea. The middle turbinate is the first major landmark encountered by the sinus surgeon.

The middle turbinate is the medial extension of the ethmoid bone. Although suspended in the nasal cavity, stability is provided by its continuation with the skull base superiorly and through the basal lamella with the lamina papyracea laterally. The middle turbinate varies greatly in its anatomical configuration. In addition, it may be pneumatized by one or more cells (concha bulla), increasing its dimensions several fold.

Access to the middle meatus is variably limited depending upon the curvature of the middle turbinate and/or the degree of pneumatization within. If only the root of the middle turbinate is pneumatized, access is frequently quite acceptable. However, if pneumatization is extensive, a reduction of the lateral portion will be required for access. Every cell within the turbinate will have an outflow tract that can be visualized.

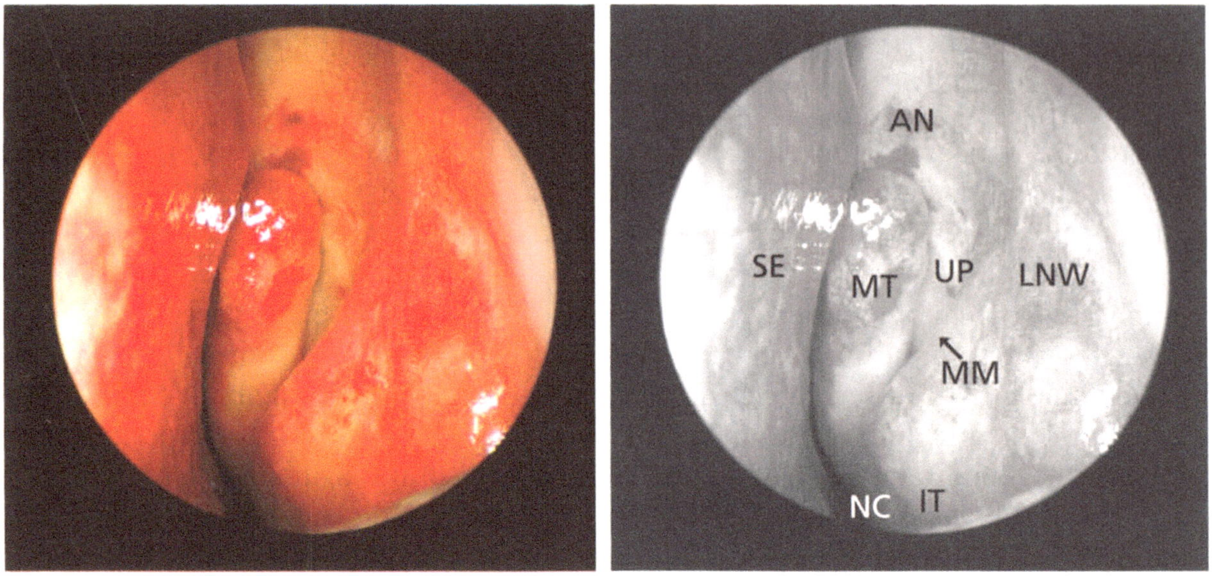

Fig. 2.1. Left side, zero degree lens, unremarkable middle turbinate with good access.

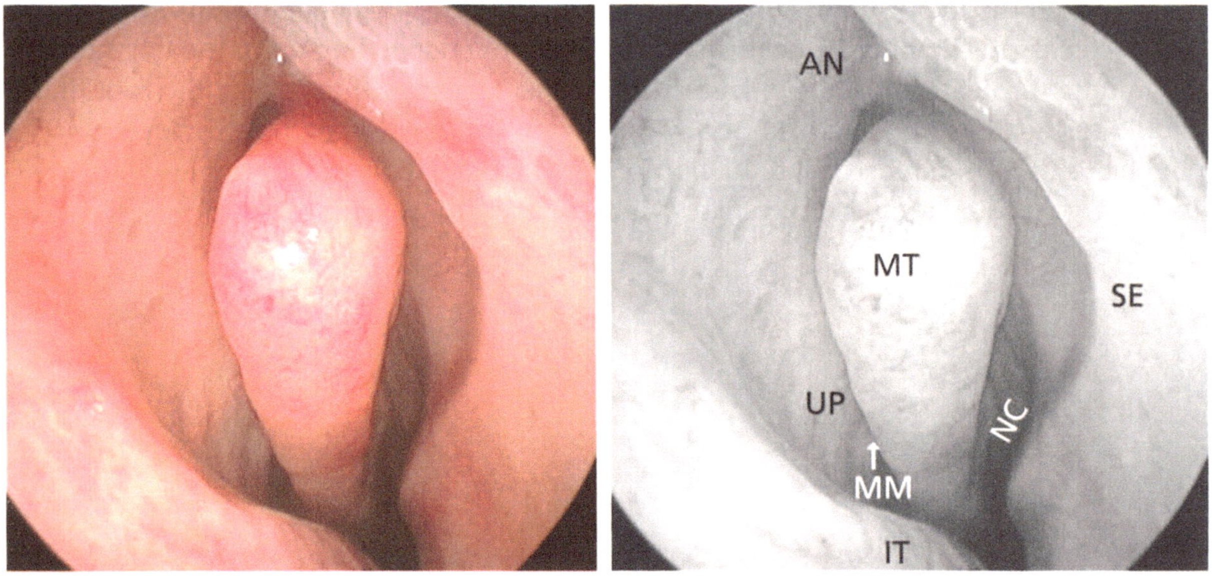

Fig. 2.2. Right side, zero degree lens, slightly bulbous middle turbinate with convex side of deviated septum.

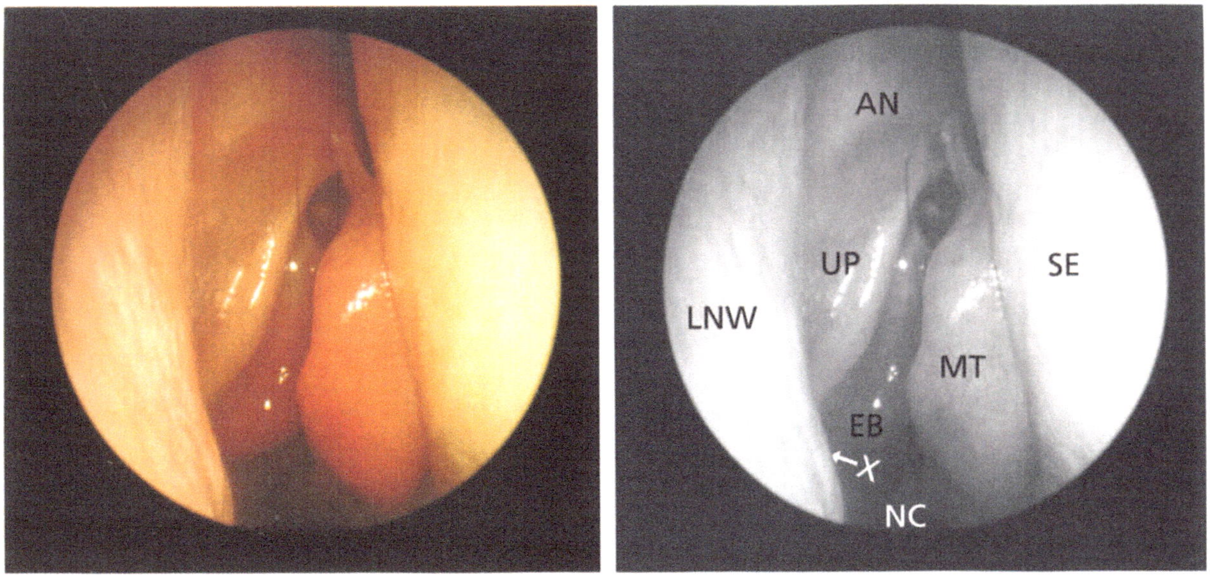

Fig. 2.3. Right side, zero degree lens, unobstructing middle turbinate, structures of middle meatus readily visible.

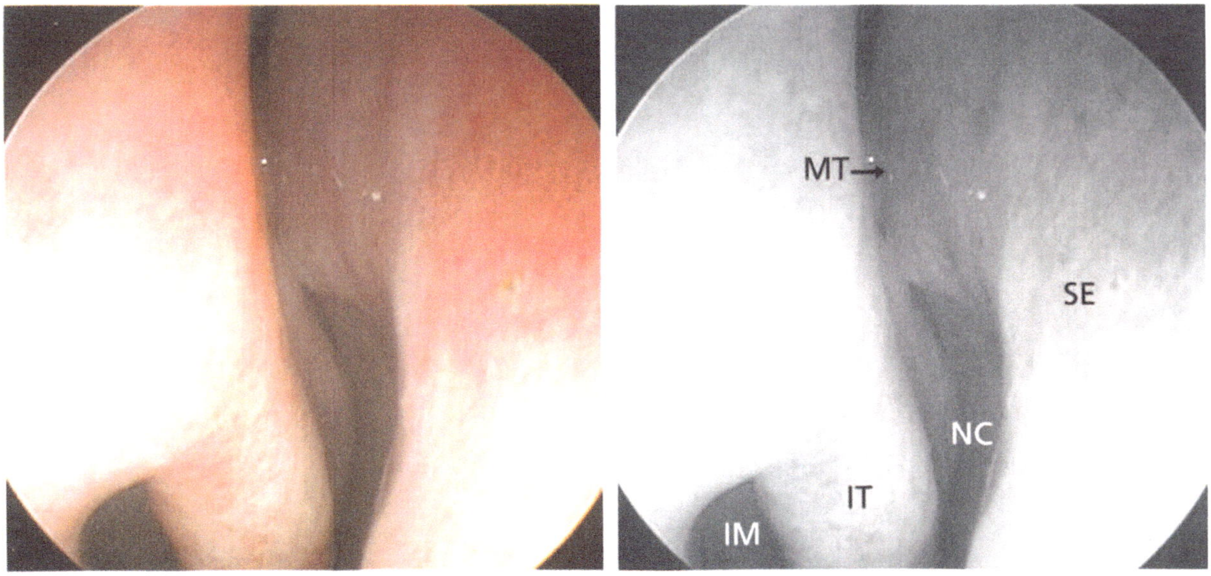

Fig. 2.4. Right side, zero degree lens, middle turbinate barely visible.

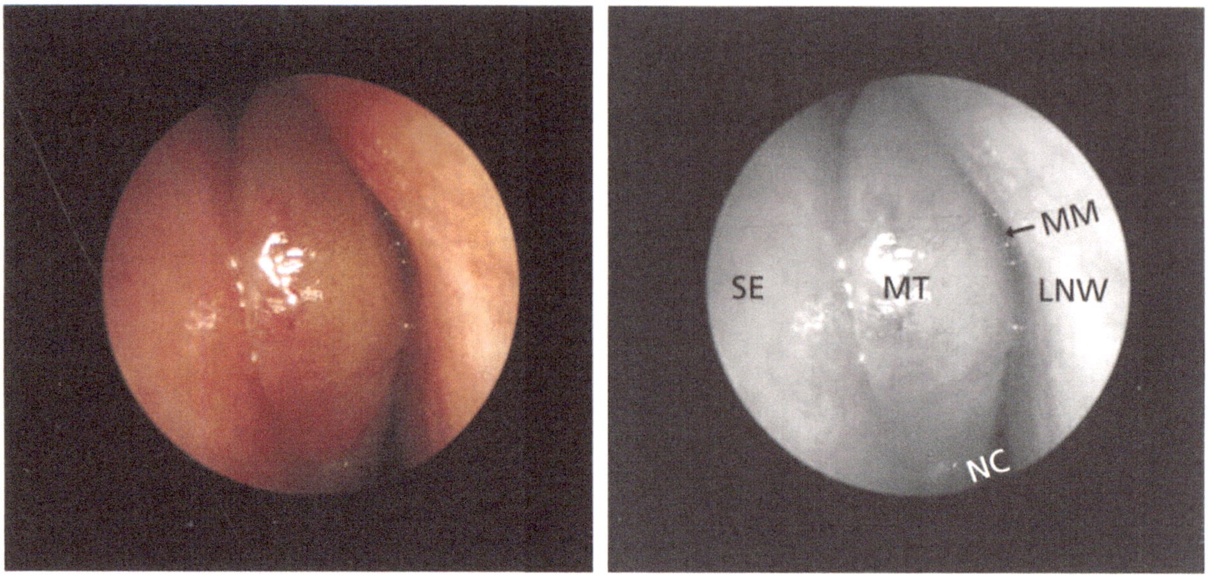

Fig. 2.5. Left side, zero degree lens, polypoid middle turbinate.

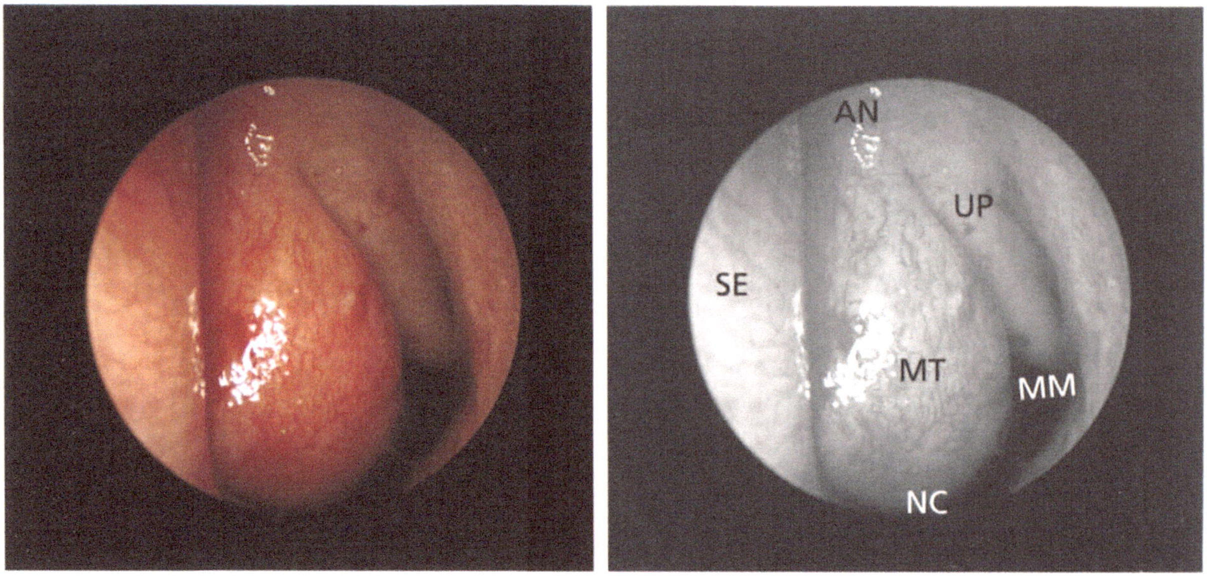

Fig. 2.6. Left side, zero degree lens, middle turbinate concha bulla, uncinate visible.

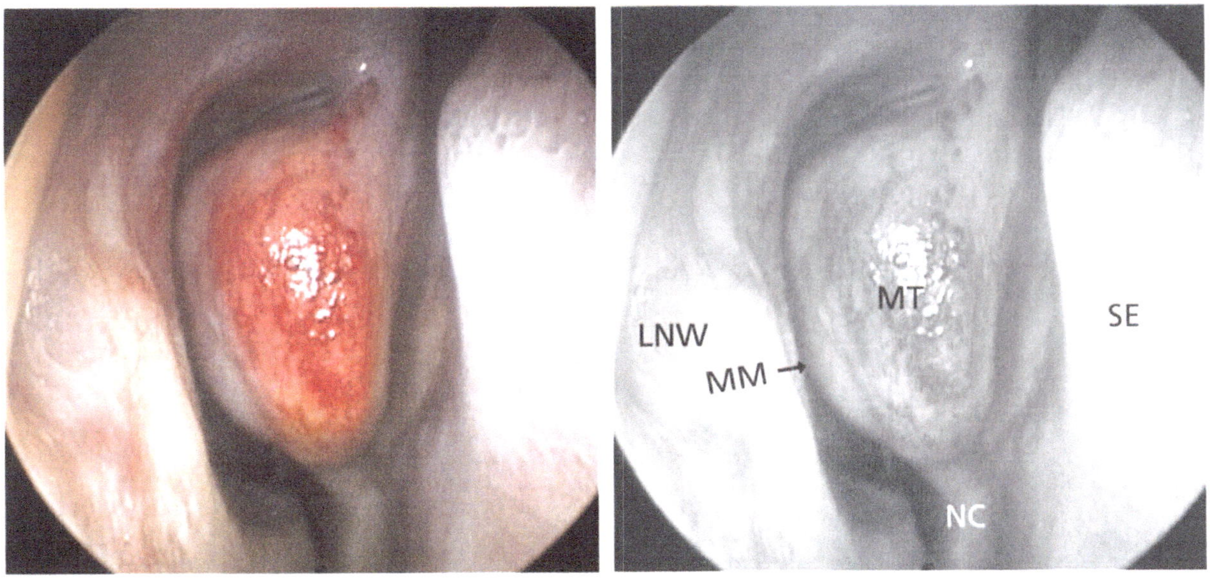

Fig. 2.7. Right side, zero degree lens, large concha bulla, middle meatus closed.

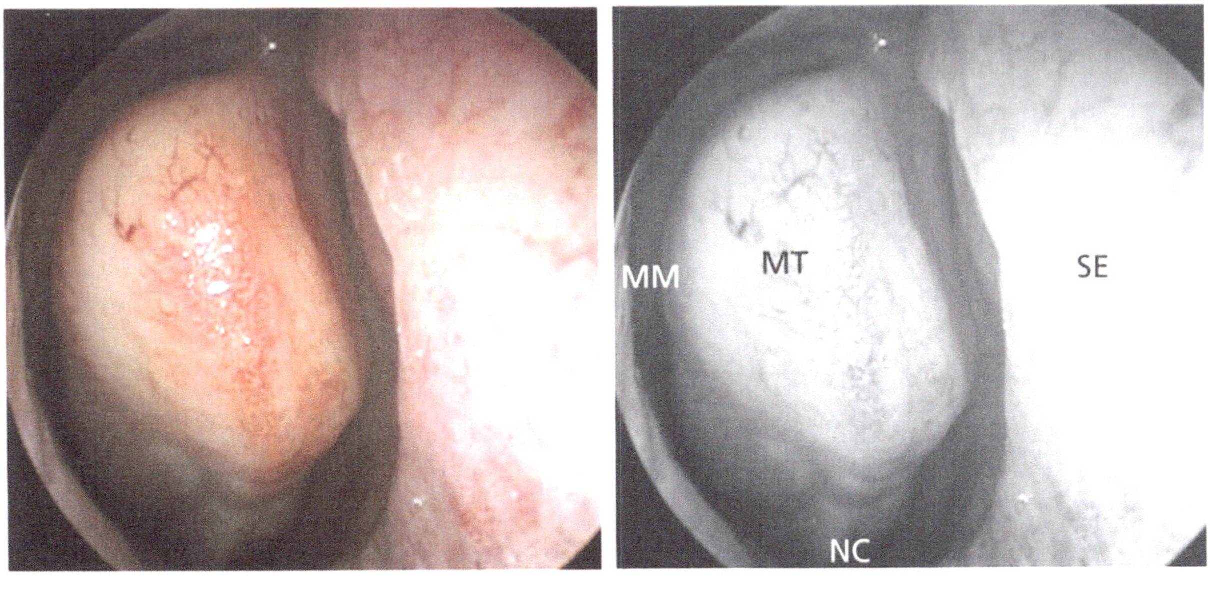

Fig. 2.8. Right side, zero degree lens, concha bulla and convexity of septal deviation.

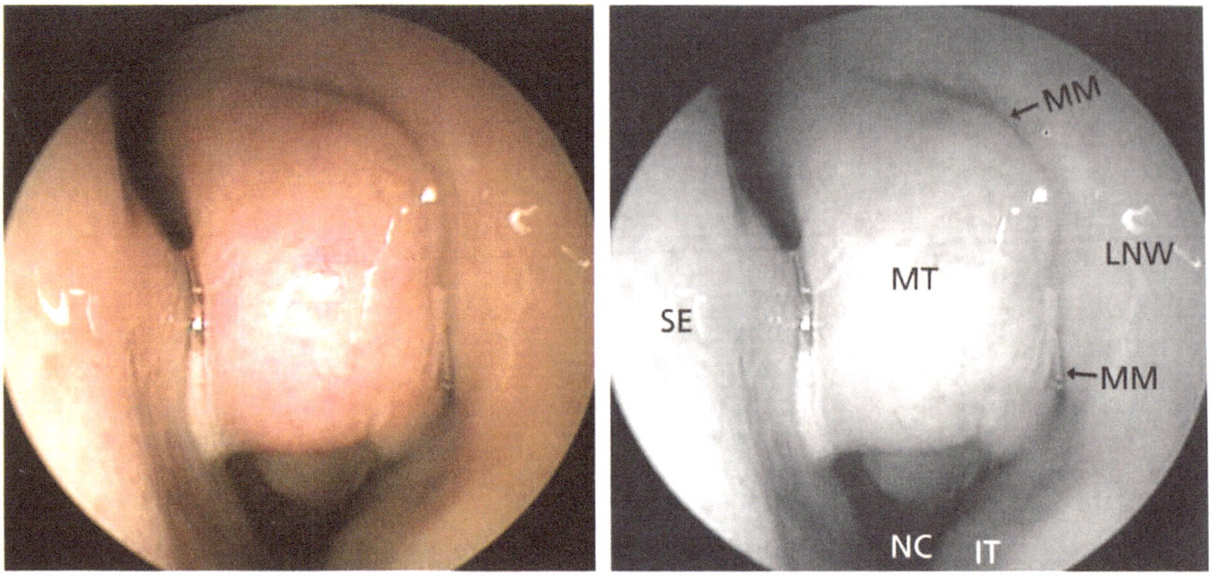

Fig. 2.9. Left side, zero degree lens, large concha bulla with polypoid change.

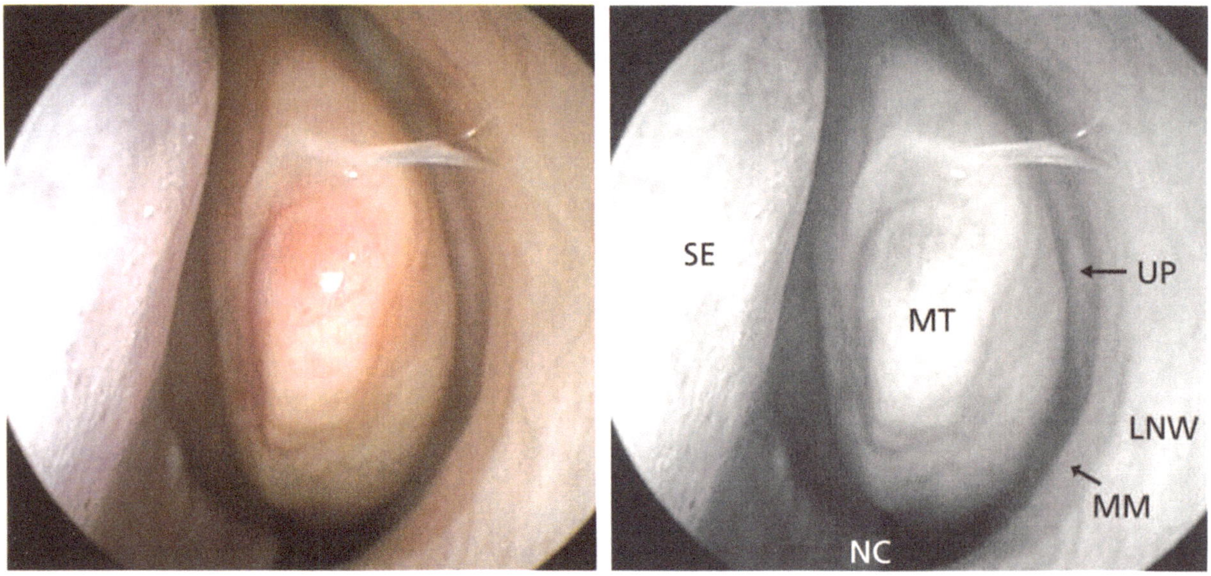

Fig. 2.10. Left side, zero degree lens, concha bulla with minimal polypoid change.

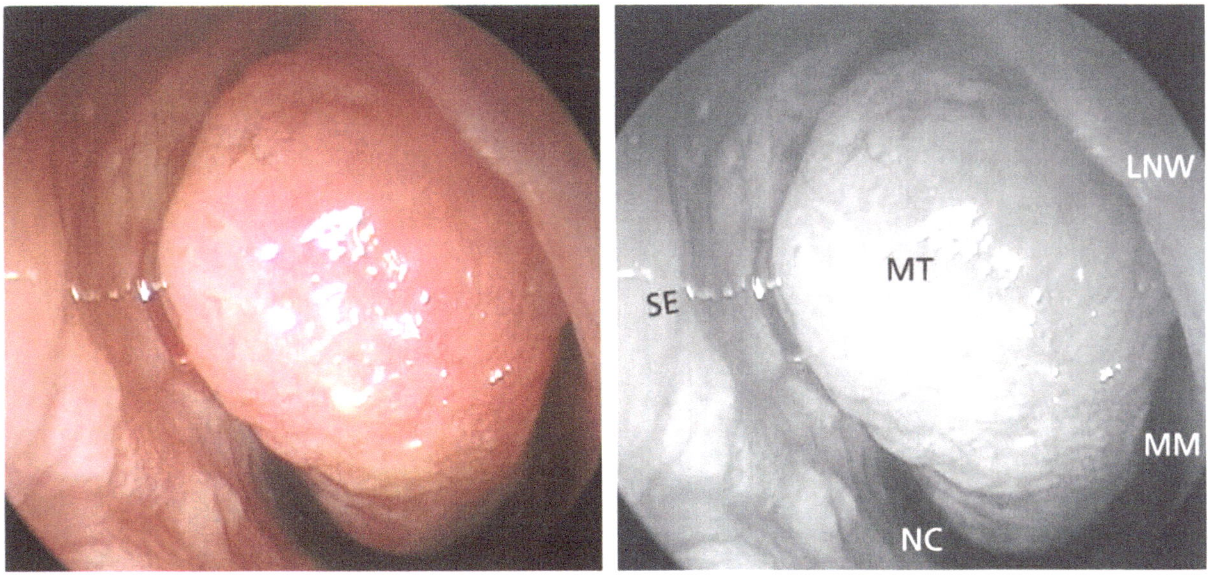

Fig. 2.11. Left side, zero degree lens, mulberry-type polypoid change with partially closed middle meatus.

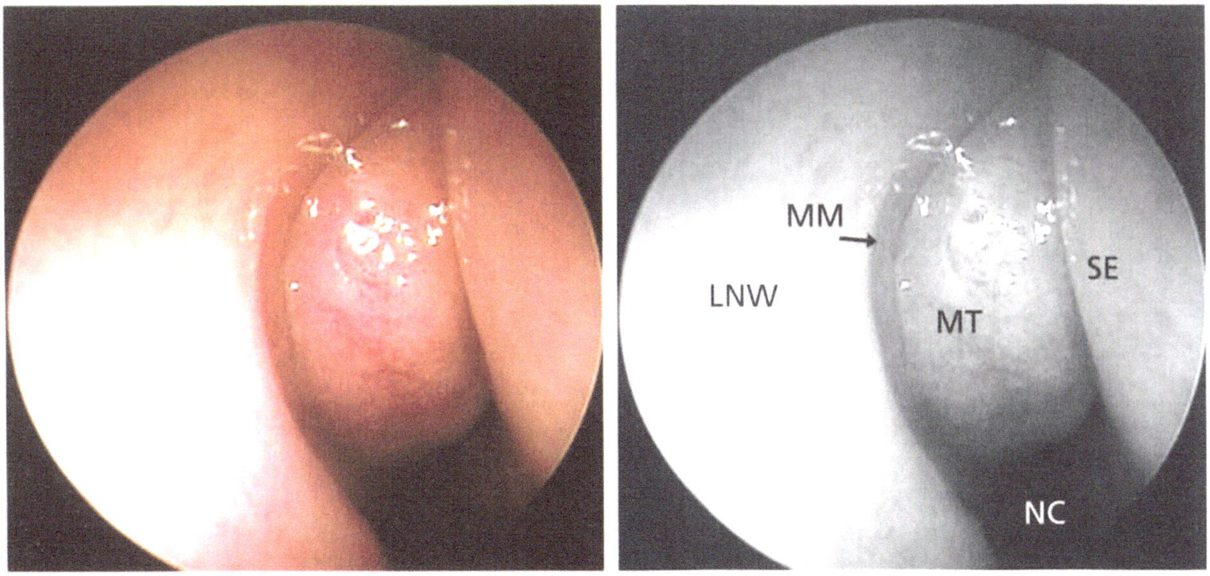

Fig. 2.12. Right side, zero degree lens, polypoid change most pronounced at interface between middle turbinate and lateral nasal wall.

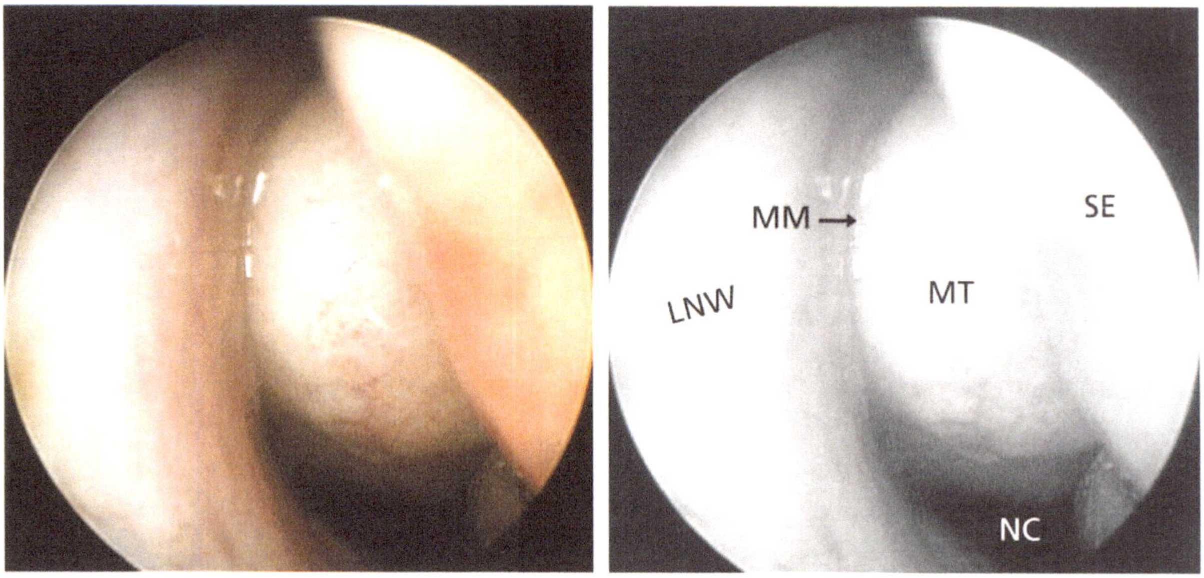

Fig. 2.13. Right side, zero degree lens, concha bulla against lateral nasal wall.

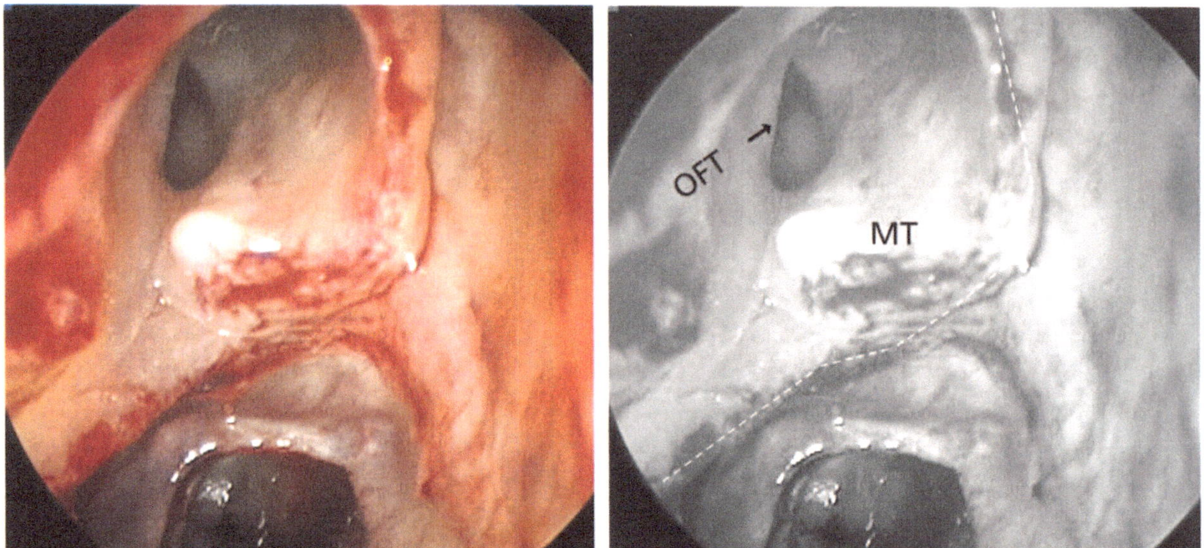

Fig. 2.14. Left side, 30 degree lens, partial resection of concha bulla with outflow tract.

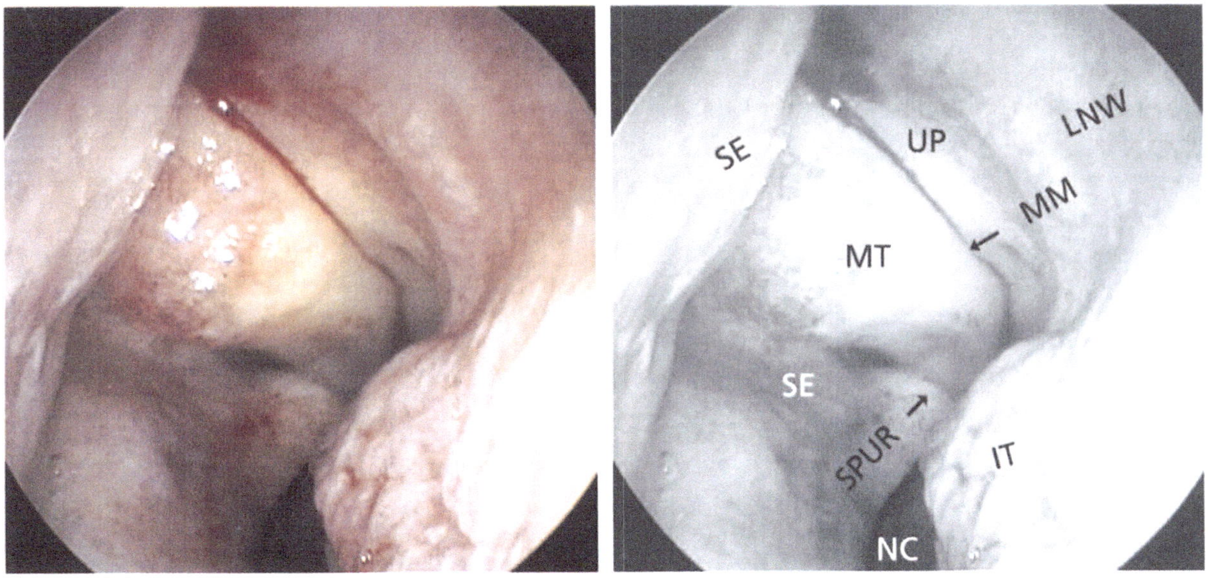

Fig. 2.15. Left side, zero degree lens, anatomical distortion secondary to rotation, and septal deviation.

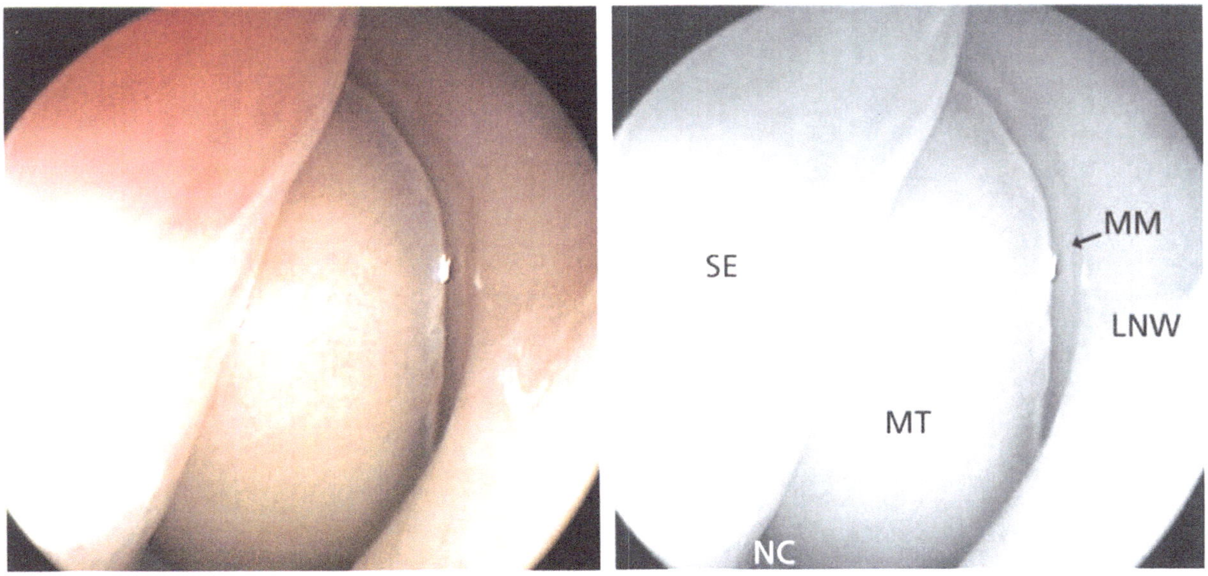

Fig. 2.16. Left side, zero degree lens, gelatinous polypoid change, most significant laterally.

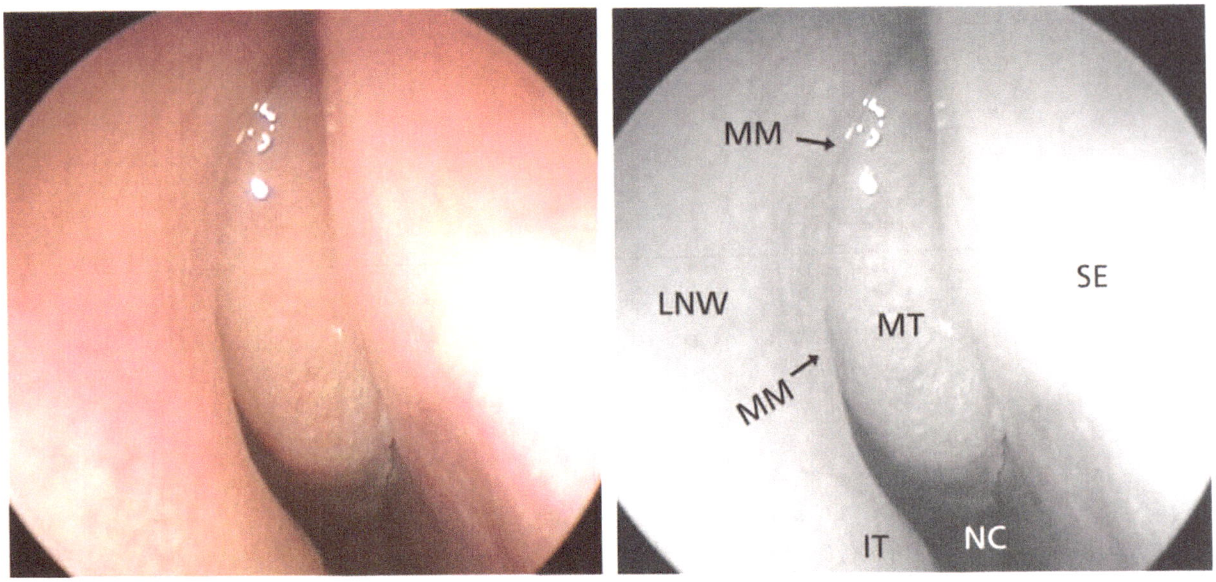

Fig. 2.17. Right side, zero degree lens, middle turbinate squeezed between the septum and the lateral nasal wall.

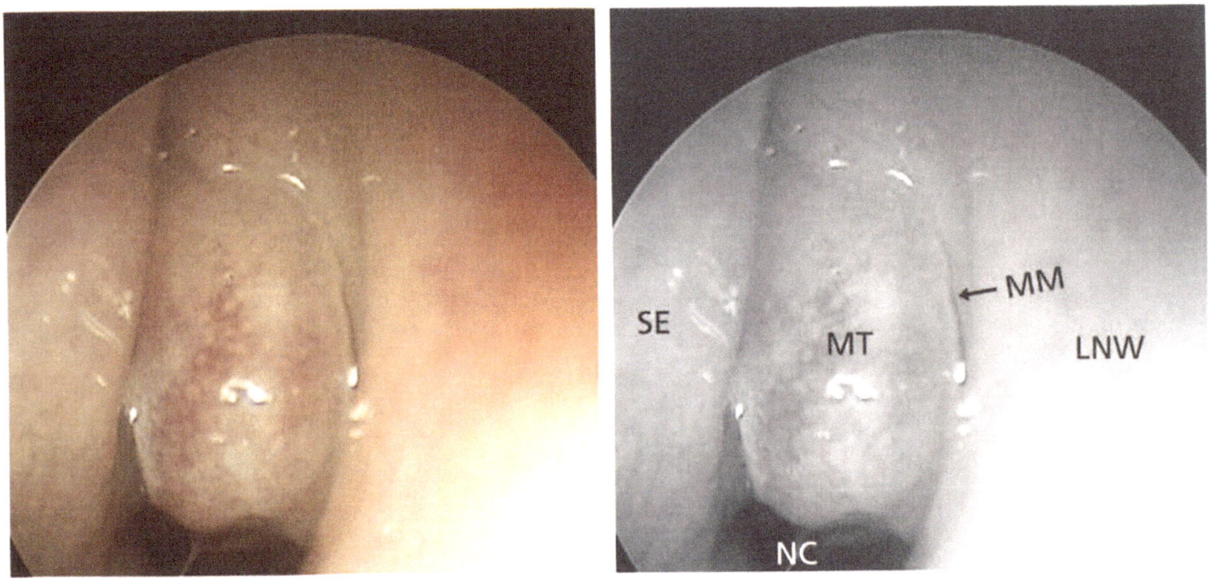

Fig. 2.18. Left side, zero degree lens, inflammation of middle turbinate with polypoid upper middle meatus.

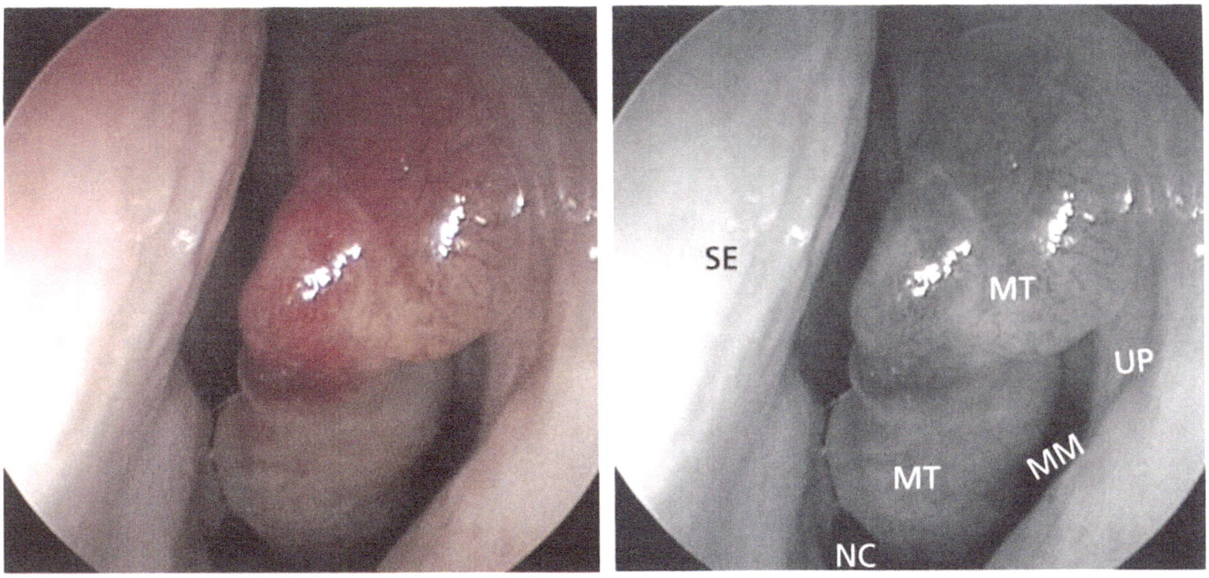

Fig. 2.19. Left side, zero degree lens, vascular polyp middle turbinate.

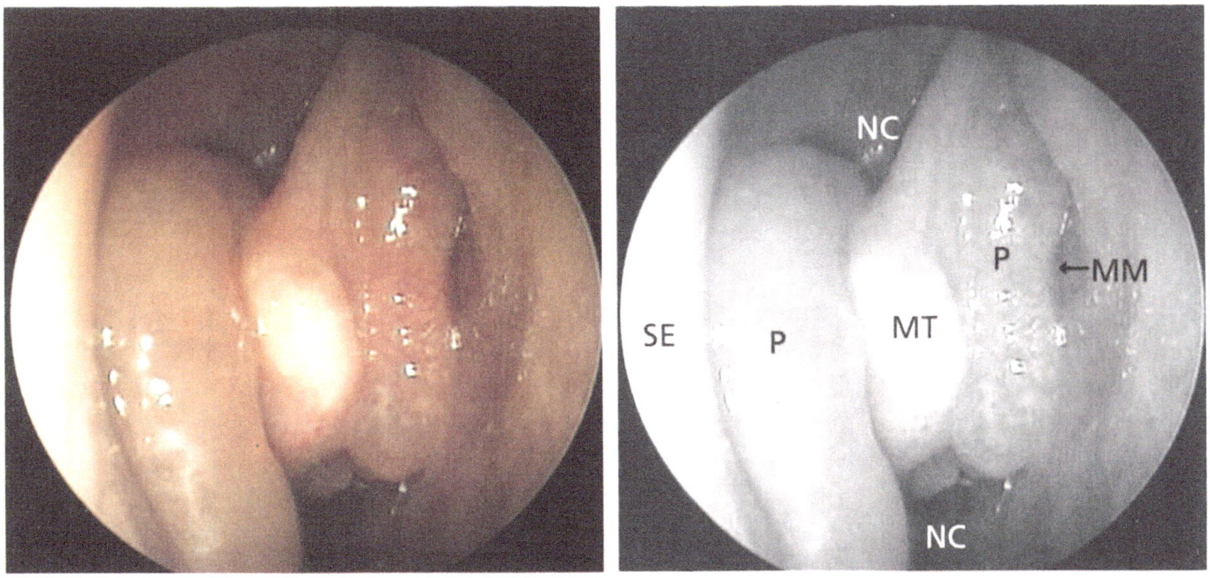

Fig. 2.20. Left side, zero degree lens, large polyp between the middle turbinate and septum with atypical polyp in middle meatus.

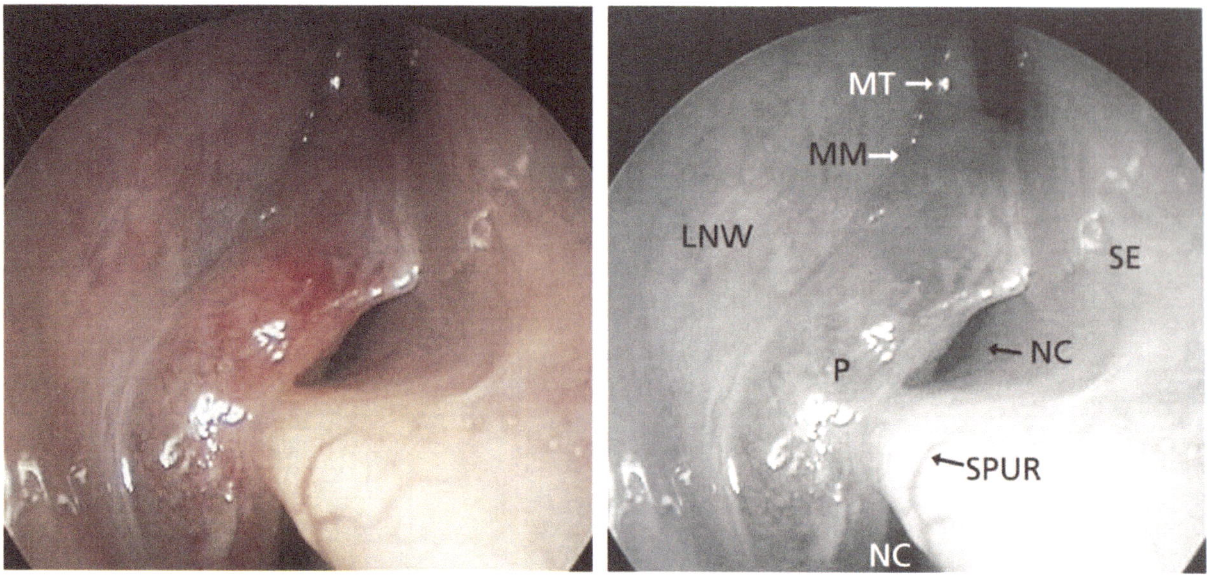

Fig. 2.21. Right side, zero degree lens, superior attachment of the middle turbinate visible at 12 o'clock, septal spur and polyp anteriorly.

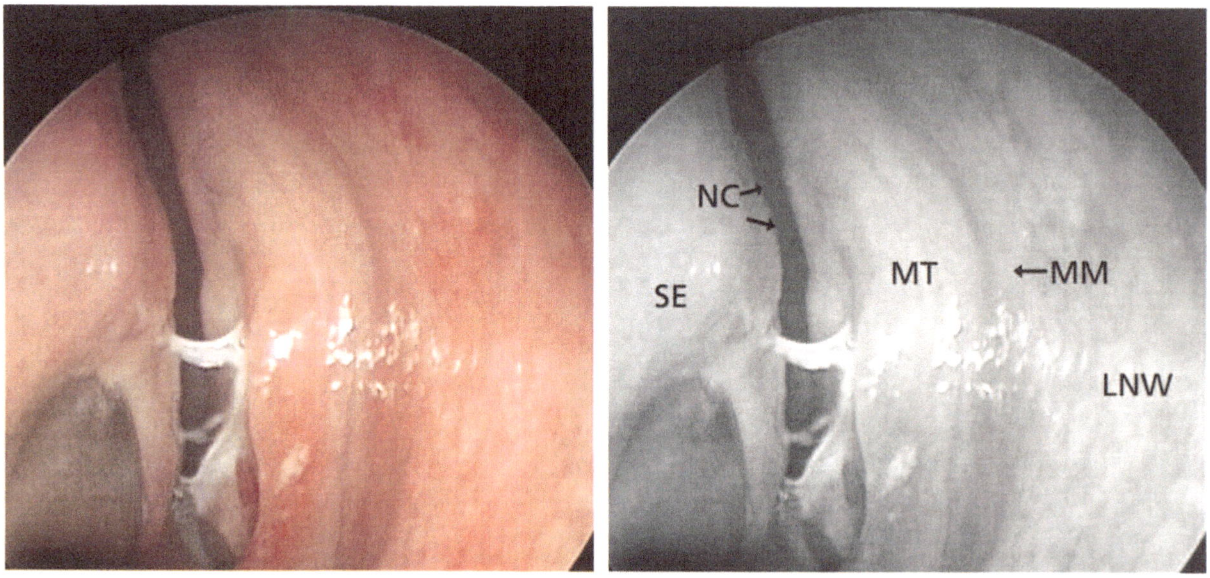

Fig. 2.22. Left side, zero degree lens, attenuation of middle turbinate with an appearance of fused middle turbinate and lateral nasal wall.

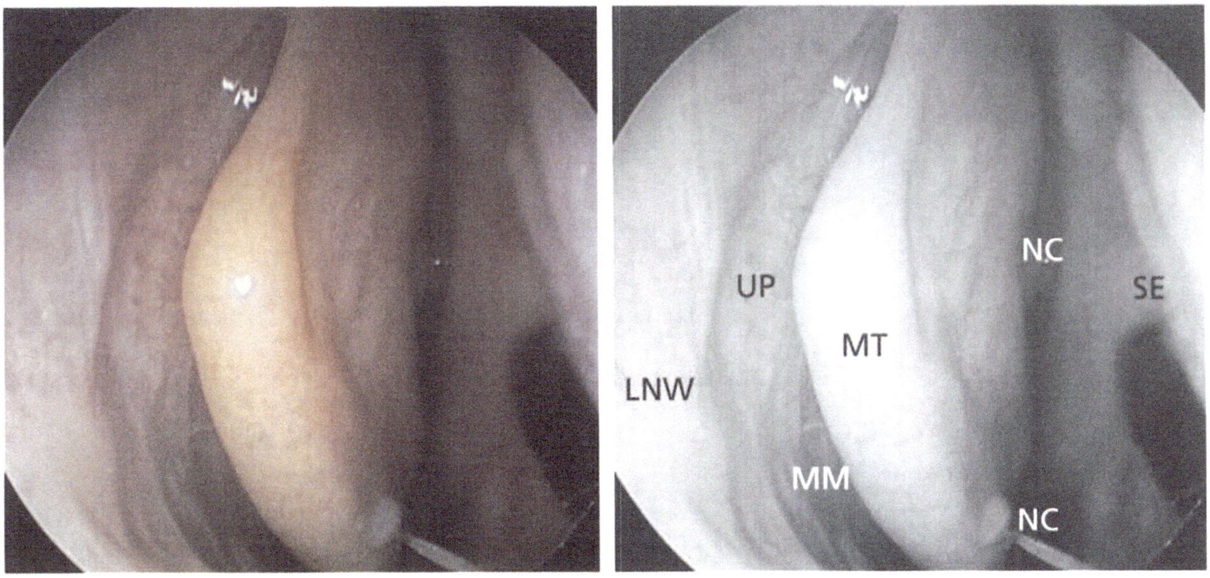

Fig. 2.23. Right side, zero degree lens, attenuation of middle turbinate secondary to septal pressure.

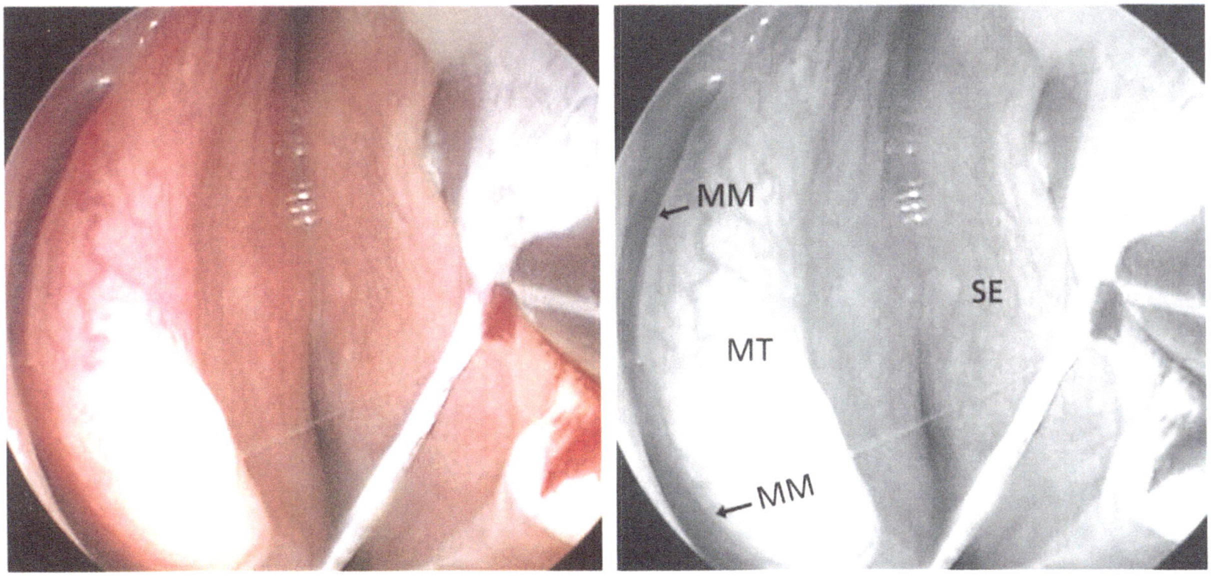

Fig. 2.24. Right side, zero degree lens, inflammatory mucosa in association with septal pressure relieved by Adson elevator. Remaining contact between middle turbinate and septum in center of picture. Septum pushed away from turbinate by Adson elevator.

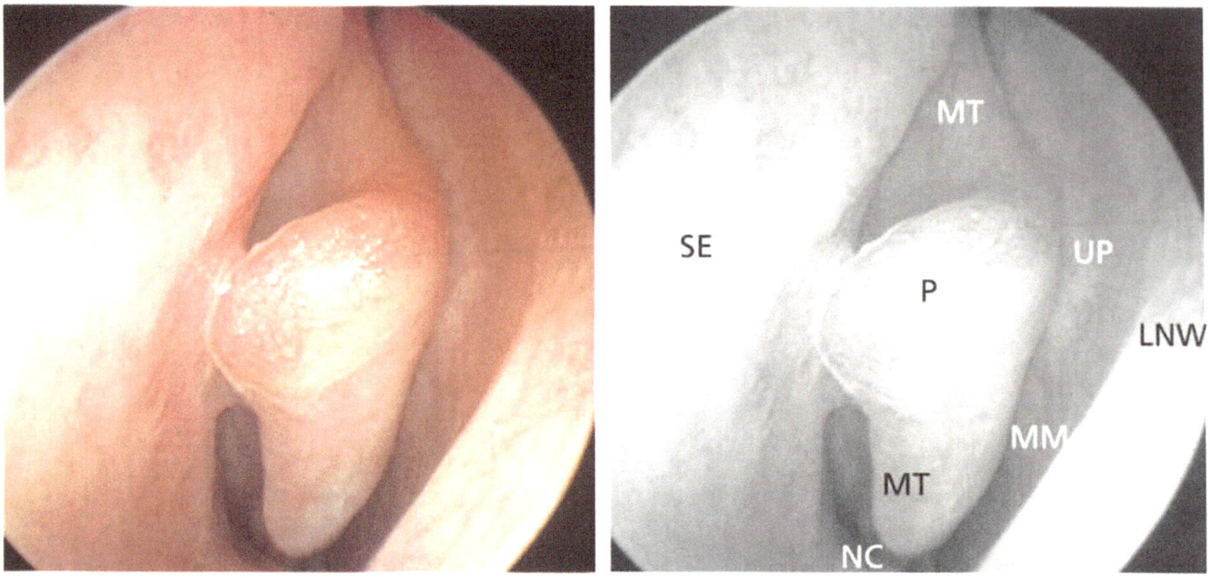

Fig. 2.25. Left side, zero degree lens, paradoxical curvature with anterior polypoid change.

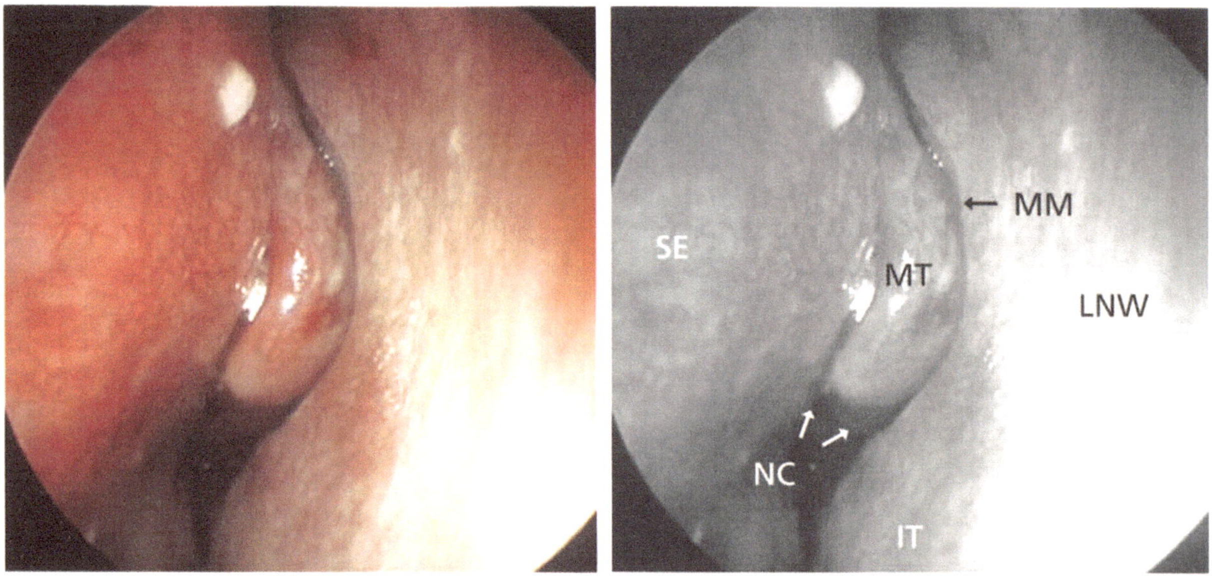

Fig. 2.26. Zero degree lens, left side paradoxical curvature and association with septal deviation and closed middle meatus.

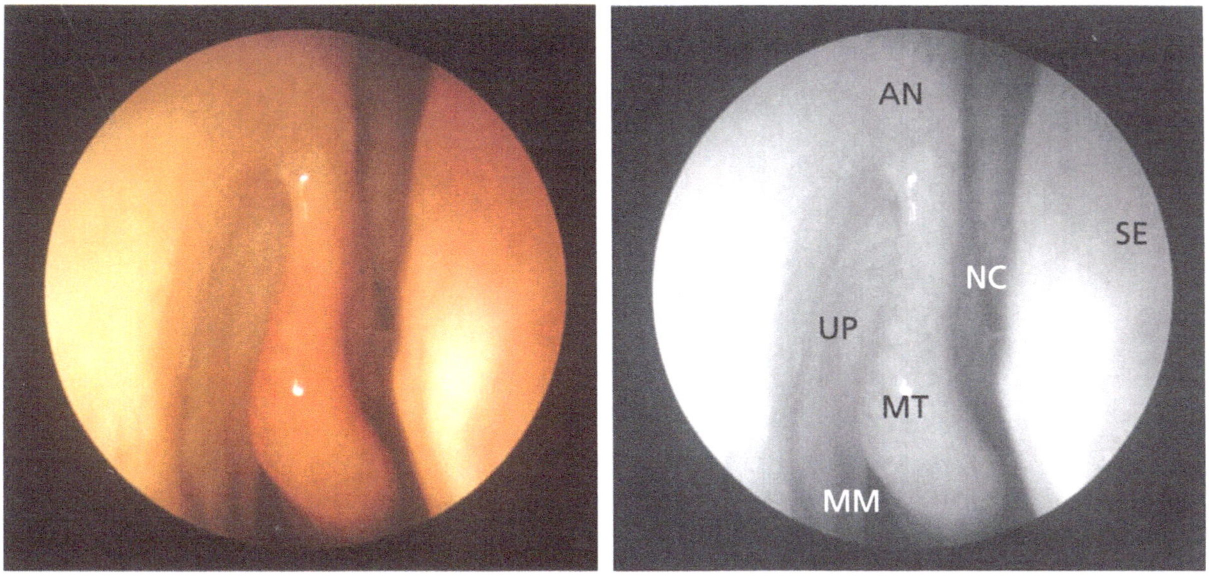

Fig. 2.27. Right side, zero degree lens, slight paradoxical curvature with visible uncinate process.

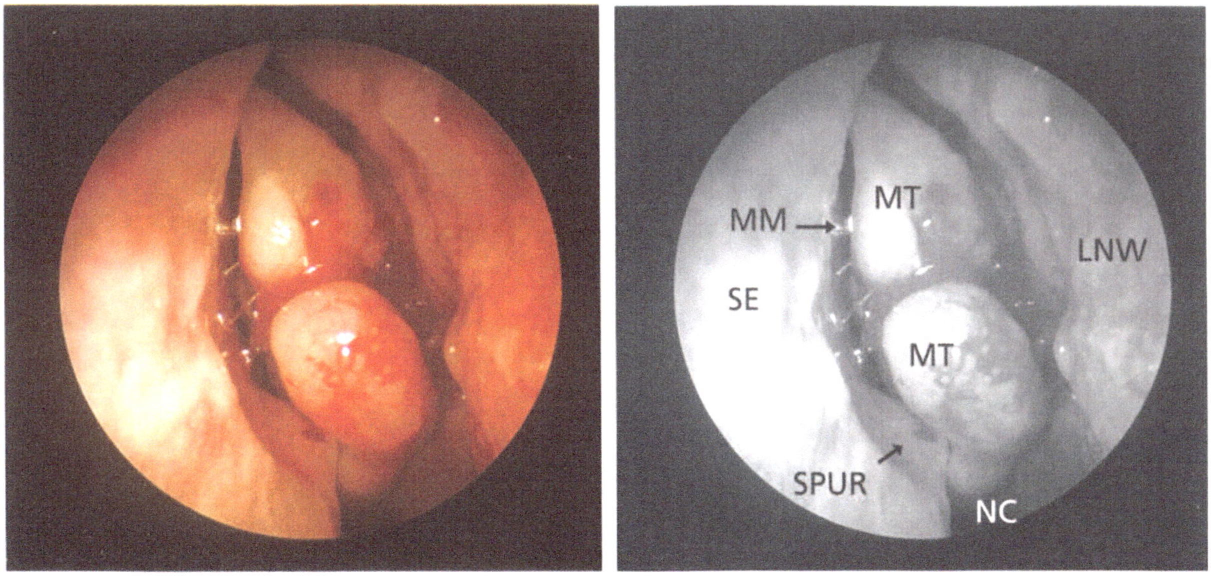

Fig. 2.28. Left side, zero degree lens, "double" middle turbinate-horizontal plane and septal spur.

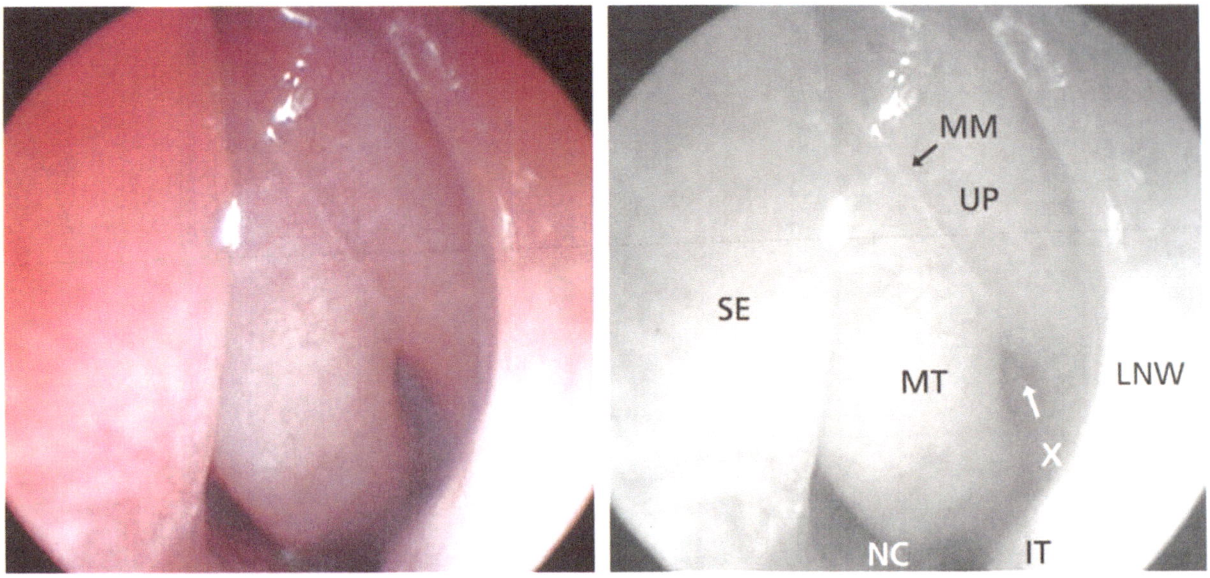

Fig. 2.29. Left side, zero degree lens, edema of middle turbinate and uncinate process.

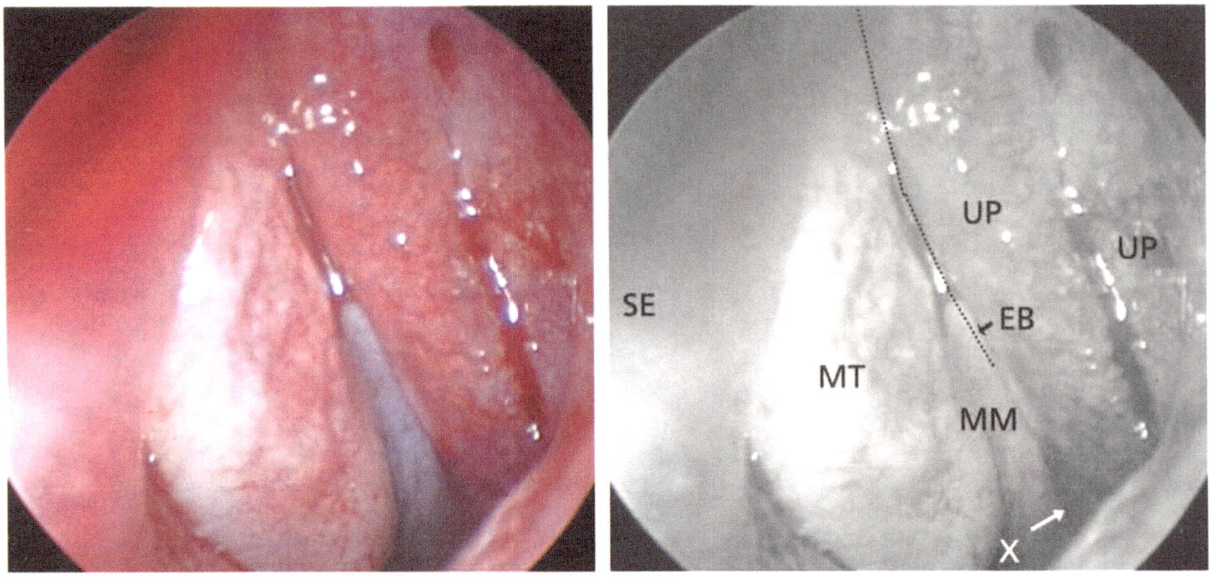

Fig. 2.30. Same as Fig. 2.29 following vasoconstriction with superior fusing of middle turbinate and "rolled" uncinate process.

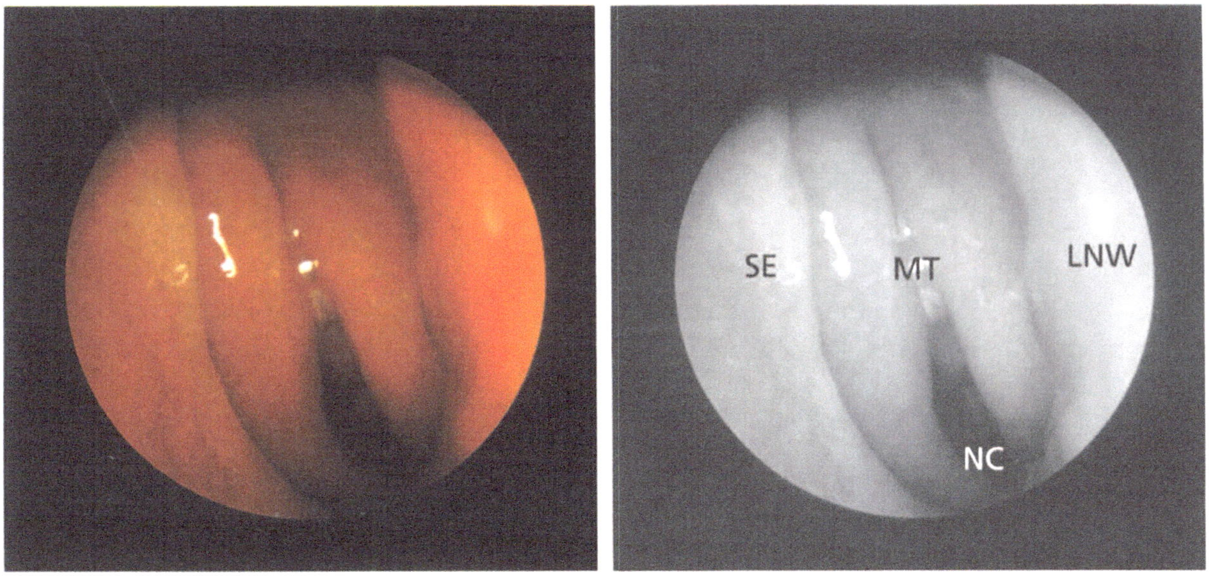

Fig. 2.31. Left side, zero degree lens, apparent duplication of middle turbinate.

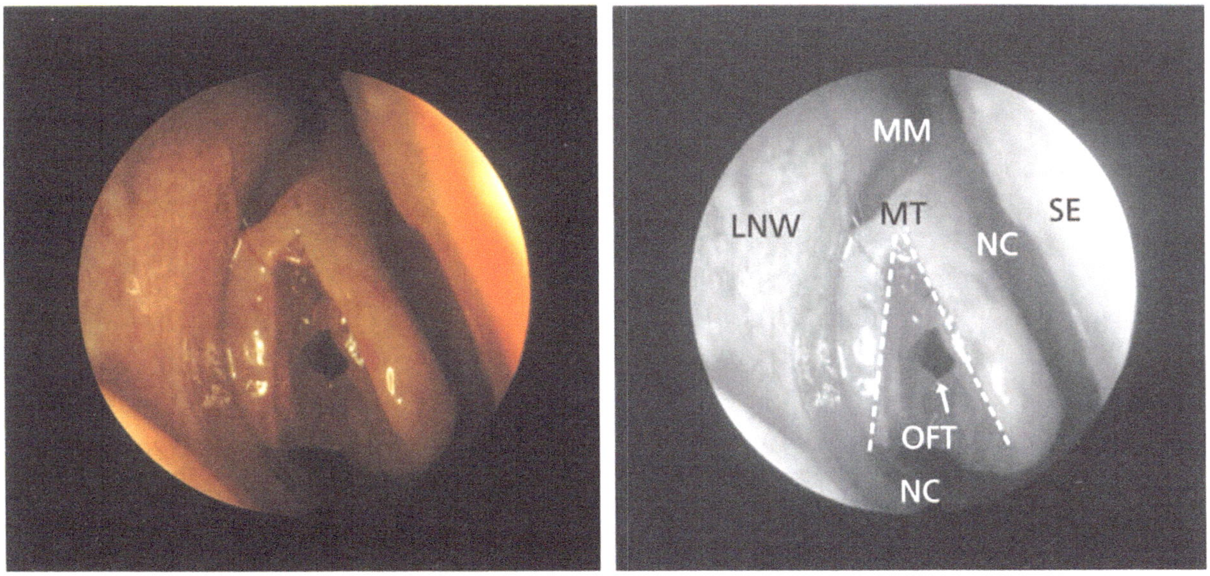

Fig. 2.32. Right side, zero degree lens, partial bi-valve of concha bulla and exposed outflow tract.

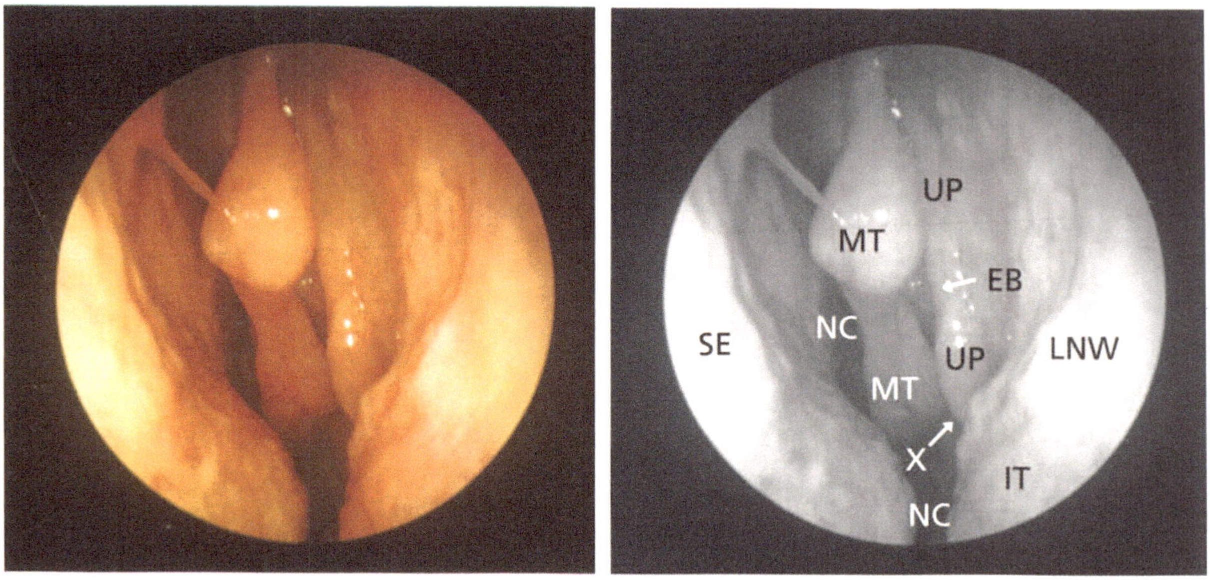

Fig. 2.33. Left side, zero degree lens, healed partial resection of middle turbinate.

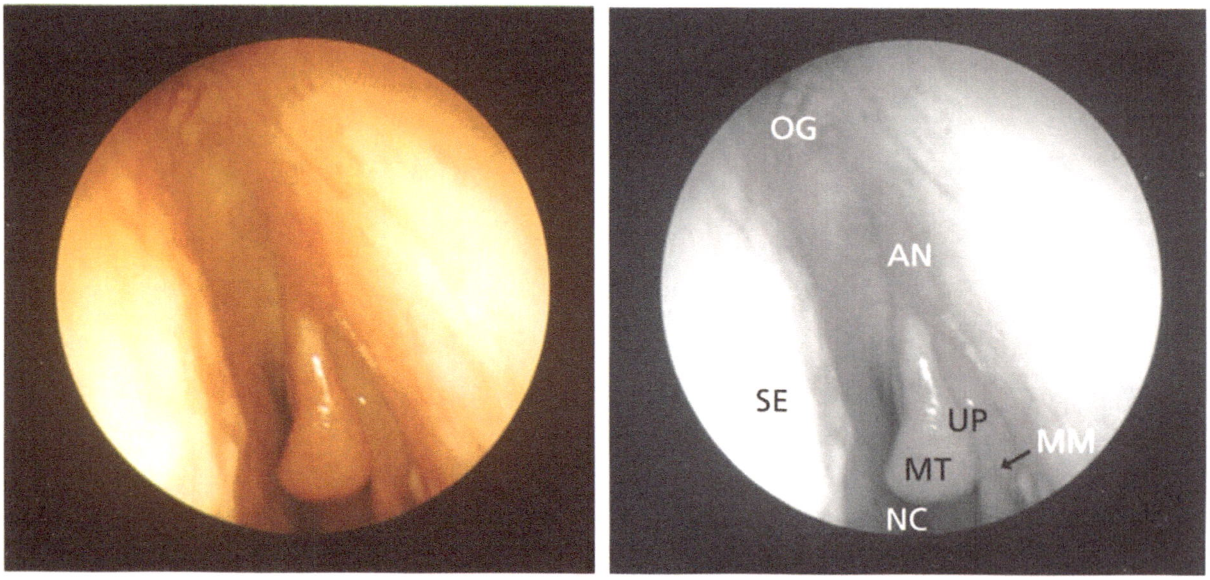

Fig. 2.34. Left side, 30 degree lens, healed partial resection of middle turbinate with olfactory groove visible.

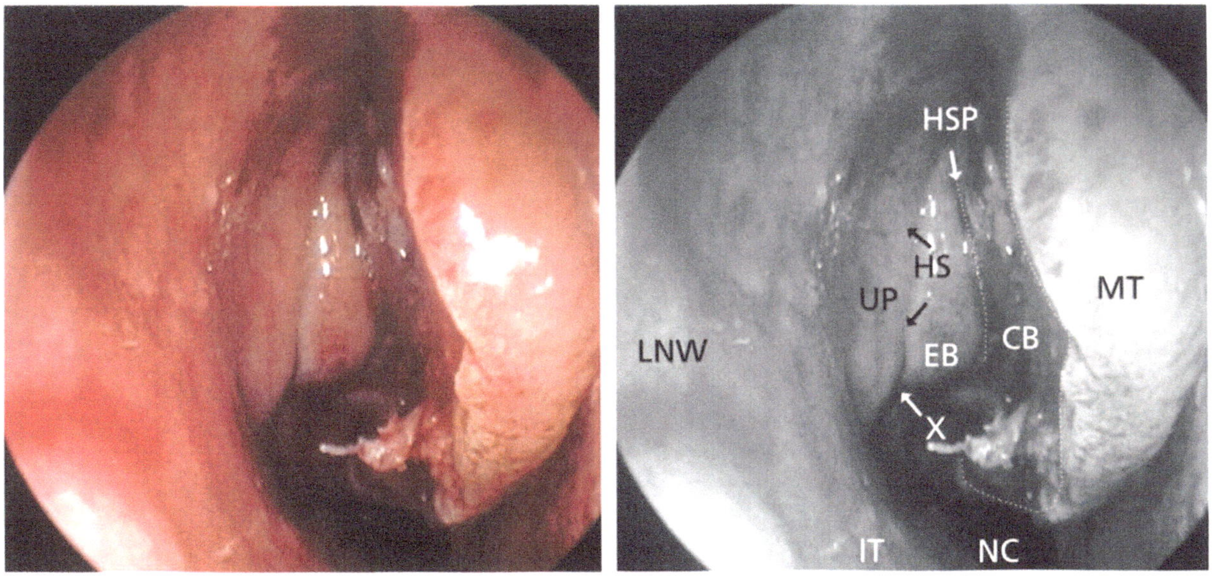

Fig. 2.35. Right side, zero degree lens, immediate post-operative partial resection of concha bulla; uncinate process and ethmoidal bulla now visible.

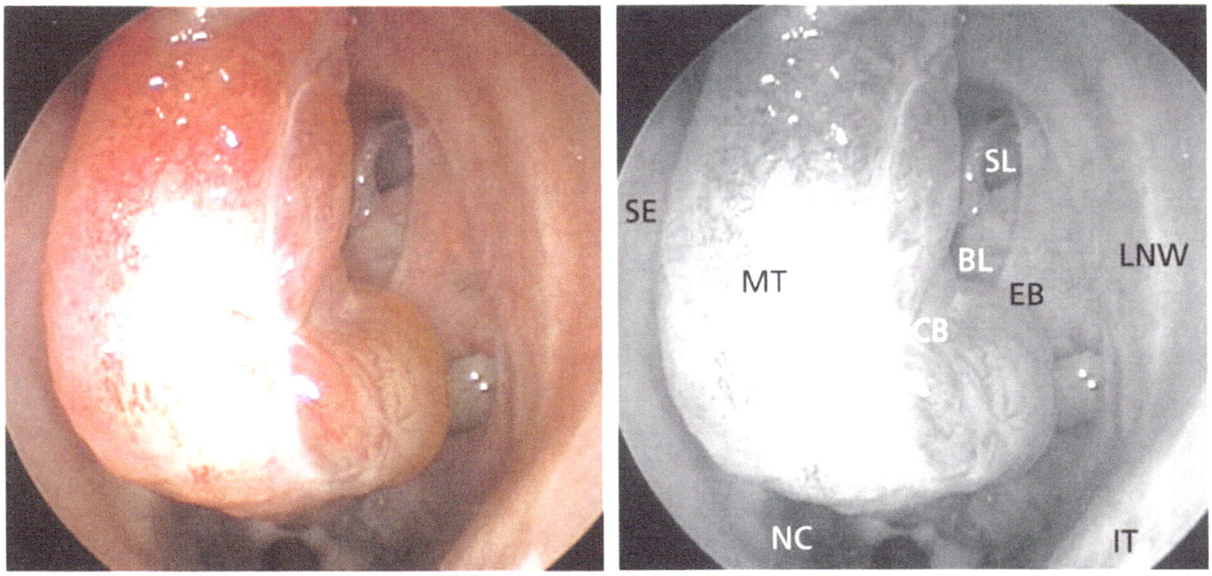

Fig. 2.36. Left side, zero degree lens, post-operative polypoid change in concha bulla from neglected outflow tract.

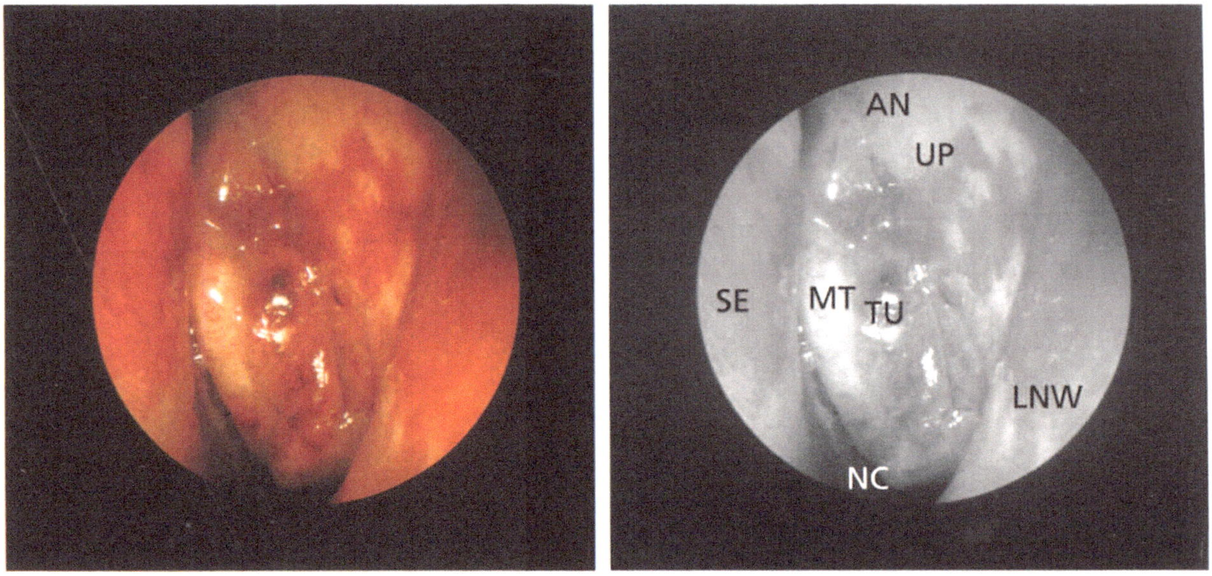

Fig. 2.37. Left side, zero degree lens, tumor of middle meatus attached to the middle turbinate.

3. The middle meatus

In health, the middle meatus is an air-containing space between the middle turbinate and lateral nasal wall. Its dimensions are largely determined by the configuration of the middle turbinate itself and the relationship between the middle turbinate and the nasal septum. The uncinate process, hiatus semilunaris, ethmoidal bulla, hiatus semilunaris posterior, horizontal and vertical portions of the basal lamella, lateral surface of the middle turbinate, exit of infundibulum, the anterior and posterior fontanelle and, depending on the insertion of the uncinate process, drainage for the frontal sinus may all be visualized by simple endoscopy of the middle meatus. The middle meatus is the most significant space in the nasal cavity when considering the drainage of the paranasal sinuses.

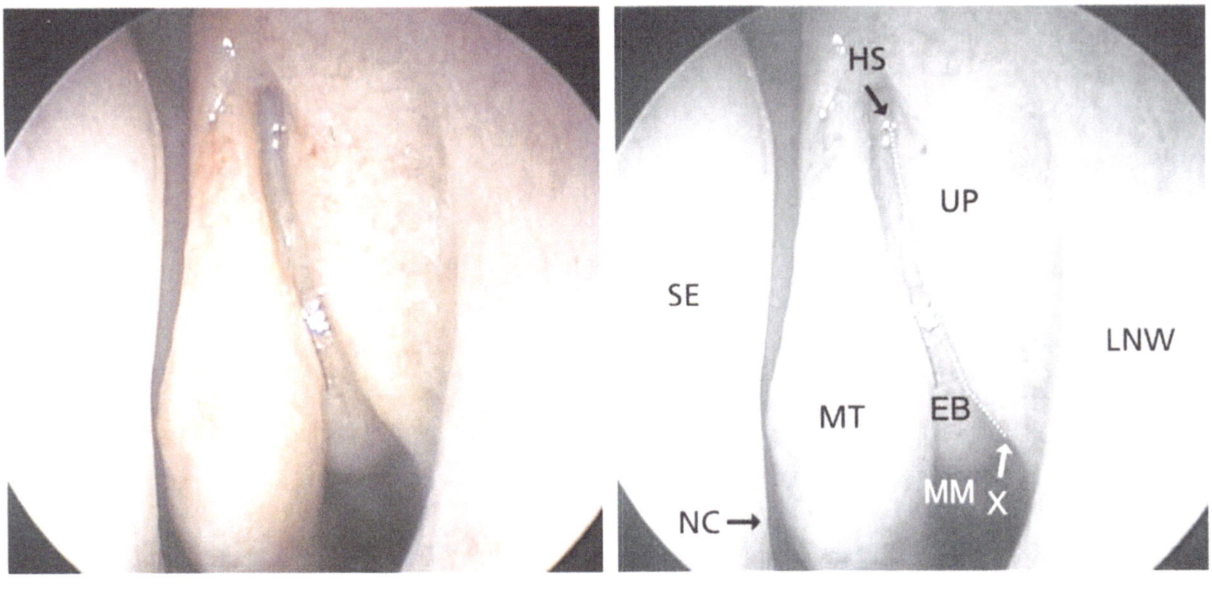

Fig. 3.1. Left side, zero degree lens, ethmoidal bulla, hiatus semilunaris, uncinate process, exit of infundibulum, and horizontal portion of basal lamella visible within the middle meatus.

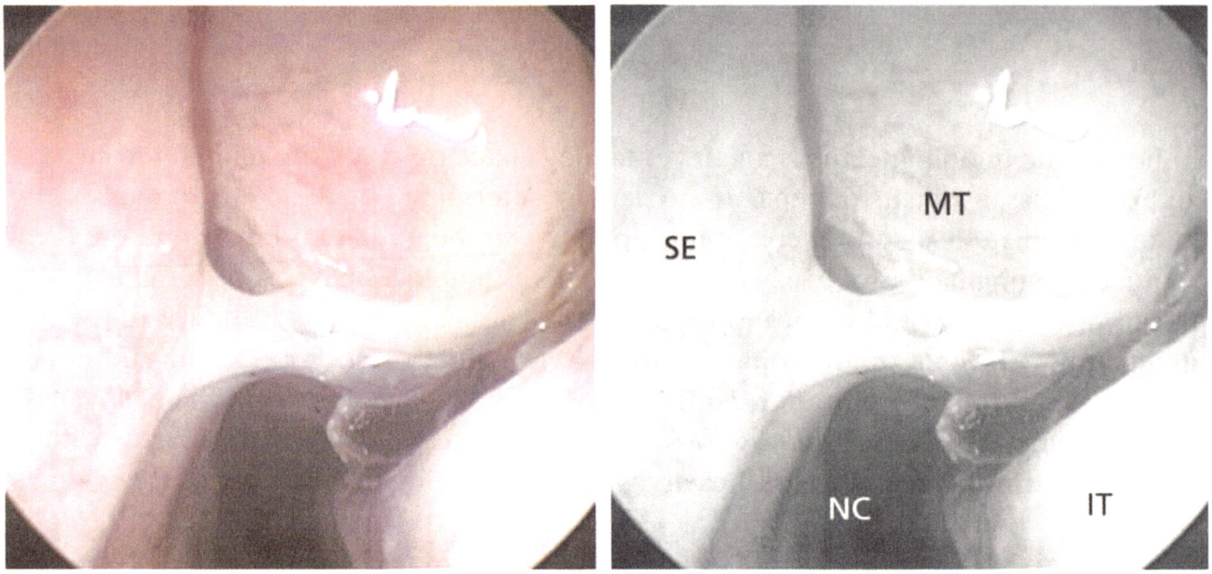

Fig. 3.2. Left side, zero degree lens, drainage from most inferior and medial portion of middle meatus between middle turbinate and inferior turbinate.

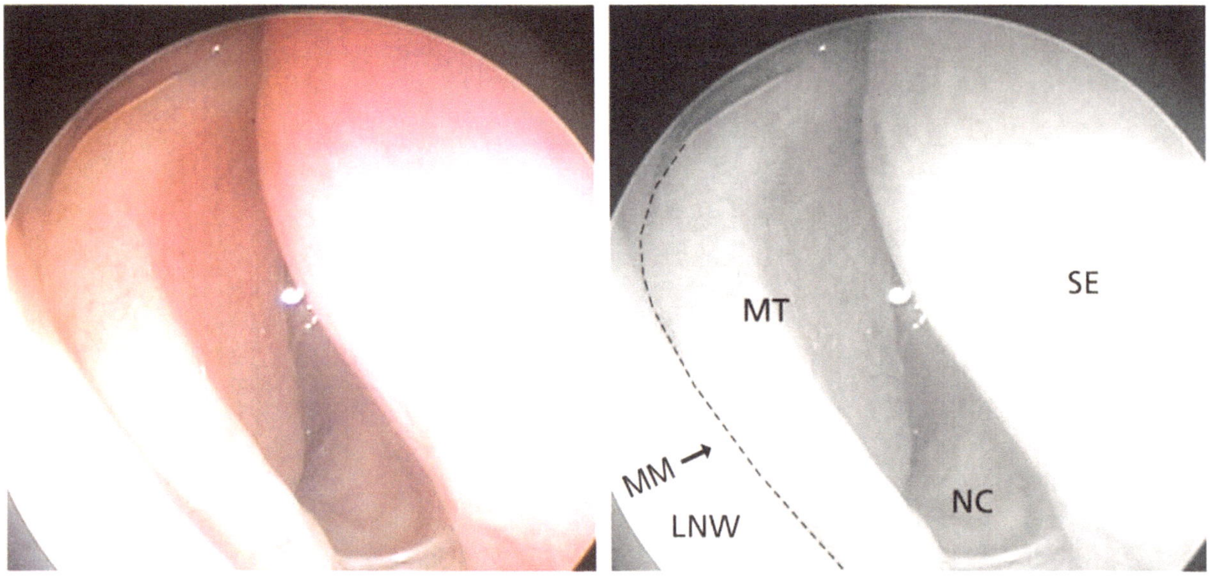

Fig. 3.3. Right side, zero degree lens, middle meatus, a potential space only due to lateralization of middle turbinate.

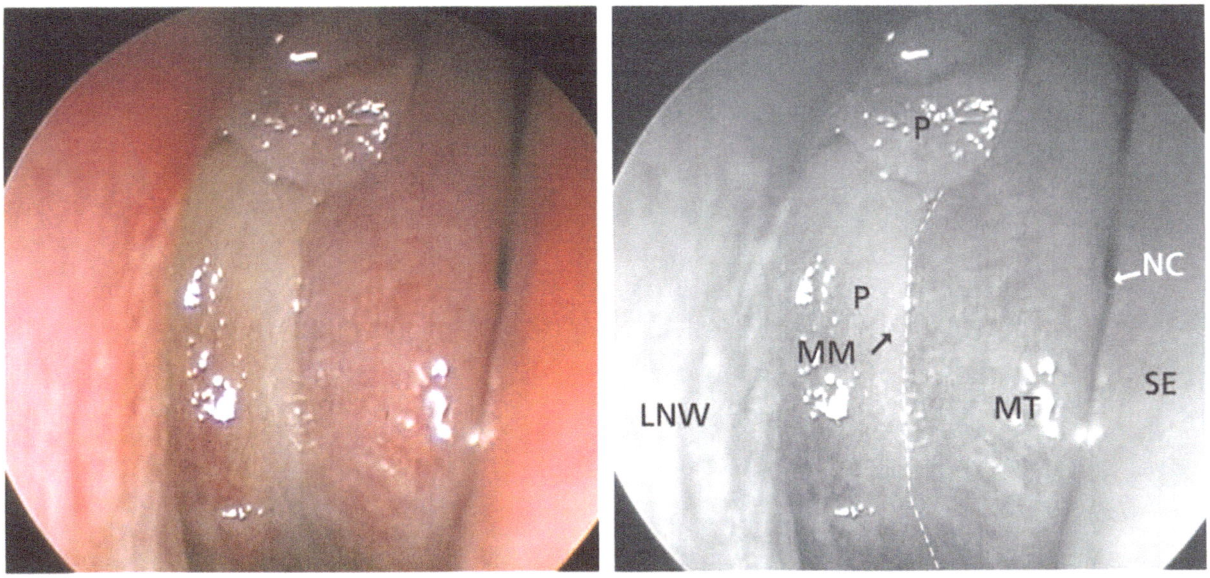

Fig. 3.4. Right side, zero degree lens, middle meatus closed by nasal polyps.

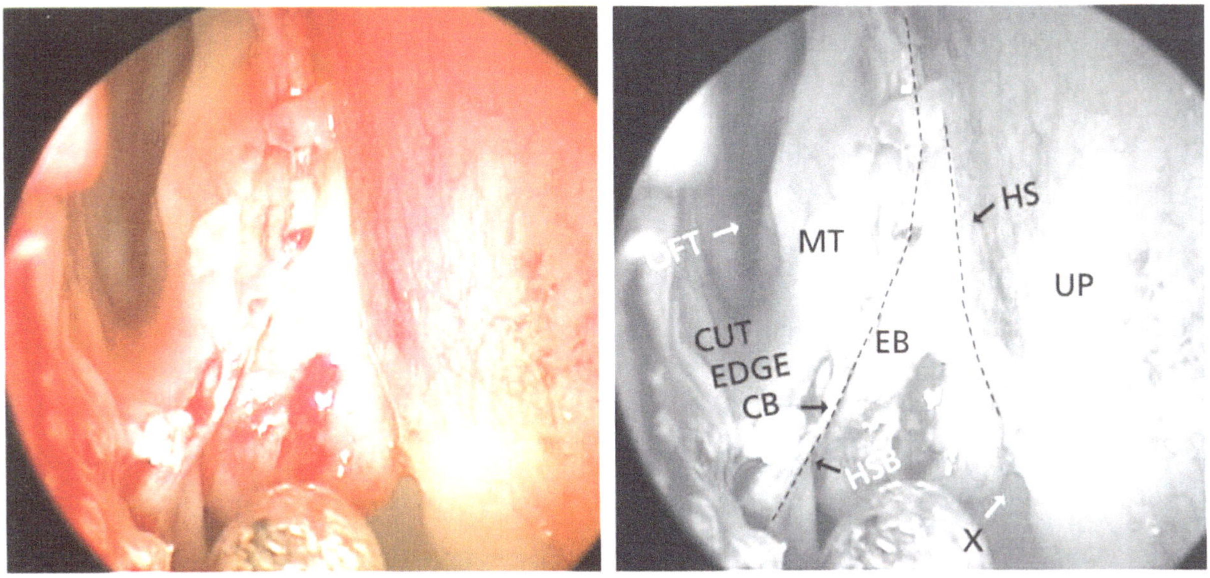

Fig. 3.5. Left side, zero degree lens, middle meatus following partial resection of concha bulla showing residual middle turbinate, outflow tract of concha, hiatus semilunaris, exit of infundibulum, uncinate process and ethmoidal bulla. Note the close contact between the lateral portion of the concha bulla and the medial and anterior portions of the ethmoidal bulla.

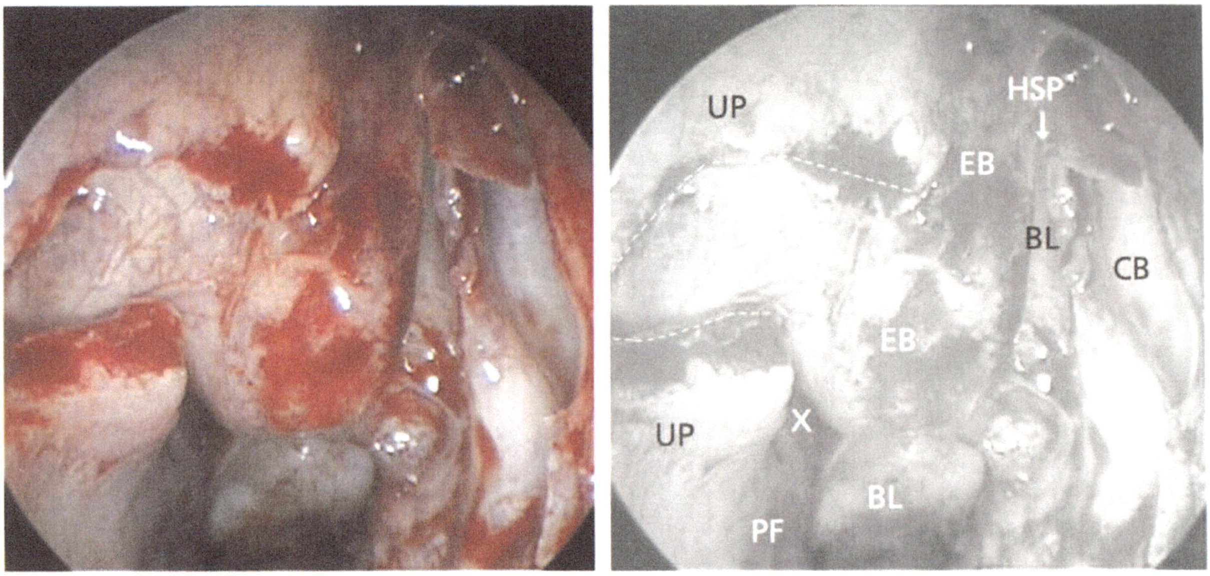

Fig. 3.6. Middle meatus following partial resection of concha bulla and partial retrograde uncinectomy showing horizontal and vertical portions of basal lamella, hiatus semilunaris posterior, ethmoidal bulla, exit of infundibulum, infundibulum itself, residual uncinate process and posterior fontanelle.

4. The uncinate process

The uncinate process, for the anatomist, is the first grand lamella. For the surgeon, it is the first structure encountered upon entering the middle meatus and the first critical anatomical landmark. The uncinate serves as the medial boundary of the infundibulum that collects drainage from the maxillary sinus, certain anterior ethmoid cells, and, in some instances, the frontal sinus. It is continuous superiorly with either the middle turbinate, base of skull, or lamina papyracea.

Anteriorly it is continuous with the lamina papyracea, marking the anterior limit of the infundibulum. The most posterior and inferior portion of the uncinate process disappears into the membranous portion of the lateral nasal wall, near the posterior fontanelle. The uncinate process is, in most instances, a relatively mobile structure which may be manipulated from either the infundibular or middle meatal side.

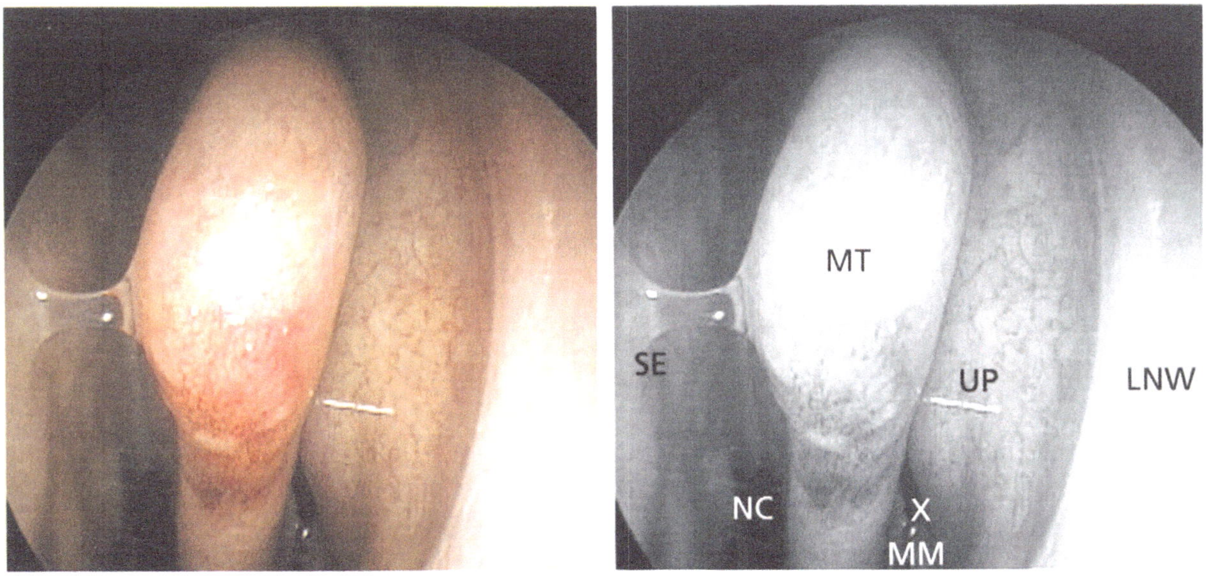

Fig. 4.1. Left side, zero degree lens, typical appearance of uncinate at endoscopy.

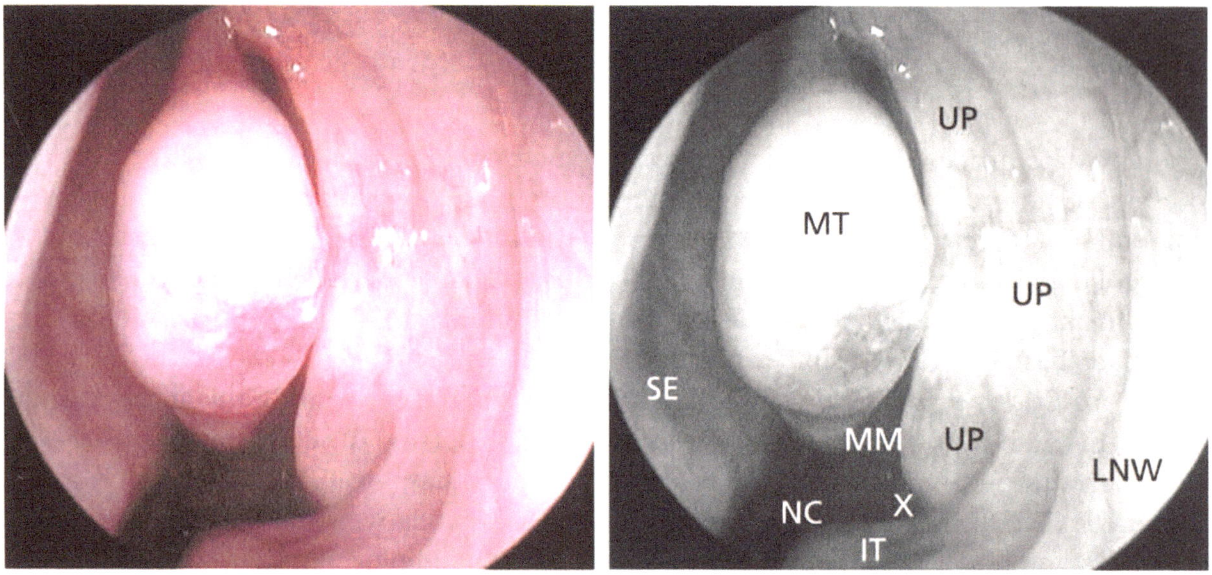

Fig. 4.2. Left side, zero degree lens, anterior displacement of uncinate process.

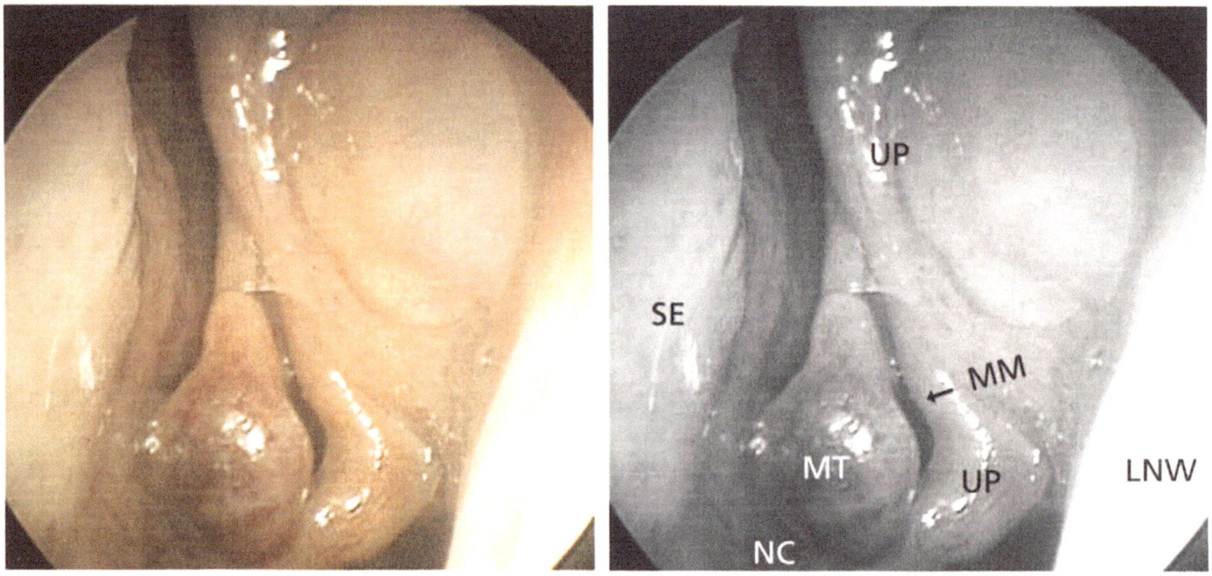

Fig. 4.3. Left side, zero degree lens, anterior displacement and distortion of uncinate process with a "rolled" edge.

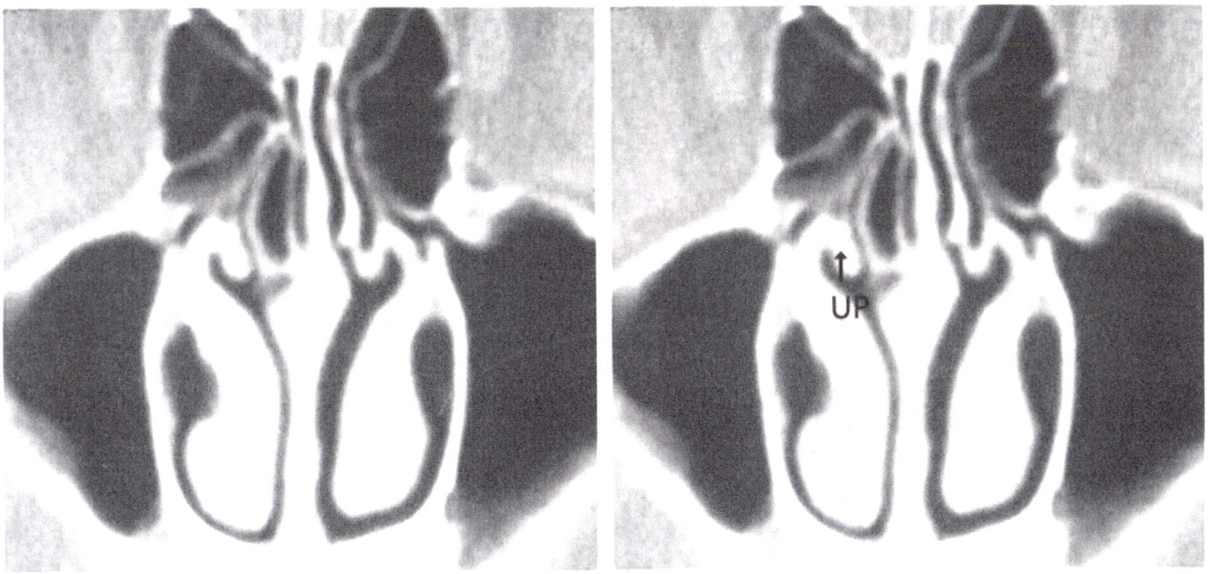

Fig. 4.4. CT showing "rolled" uncinate.

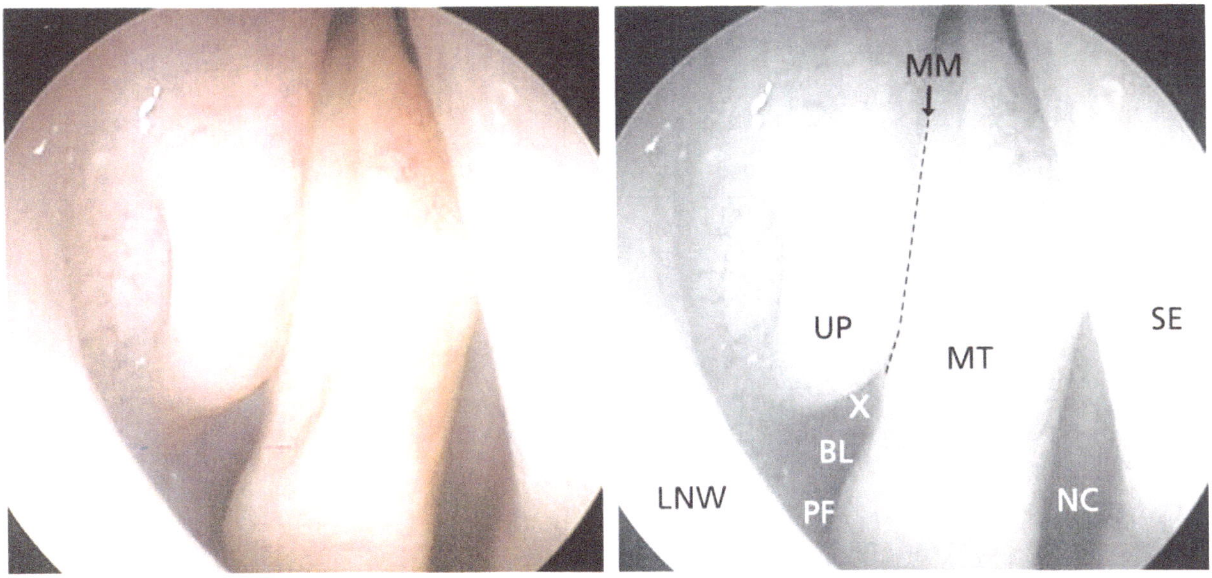

Fig. 4.5. Right side, zero degree lens, another "rolled" uncinate with unusual middle turbinate configuration.

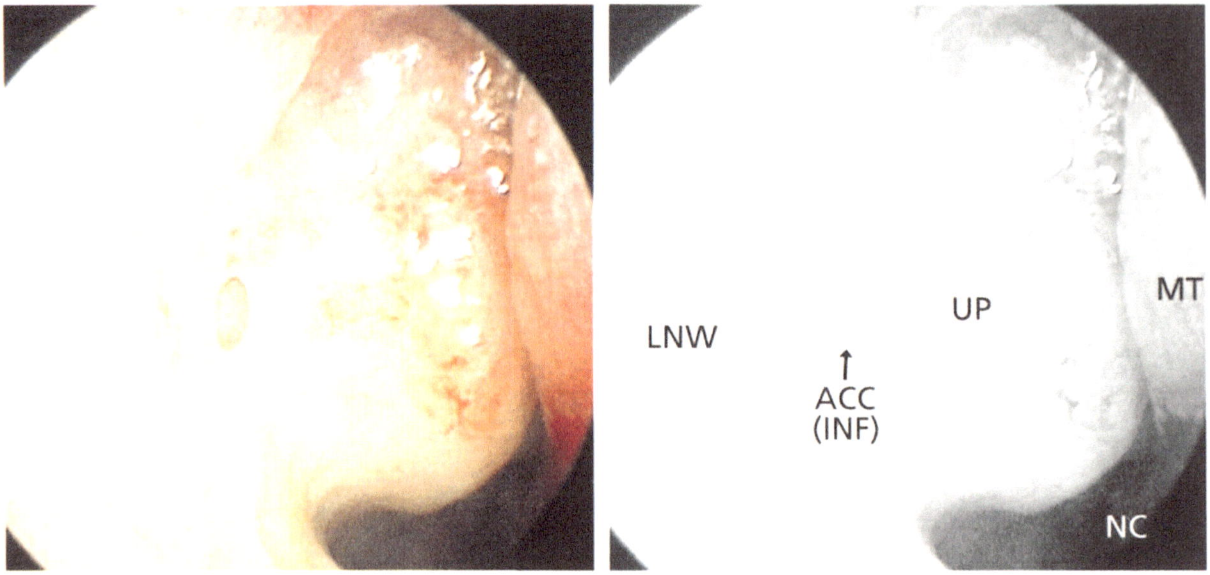

Fig. 4.6. Right side, zero degree lens, accessory entrance to infundibulum through the uncinate process; uncinate process itself completely covers the ethmoidal bulla.

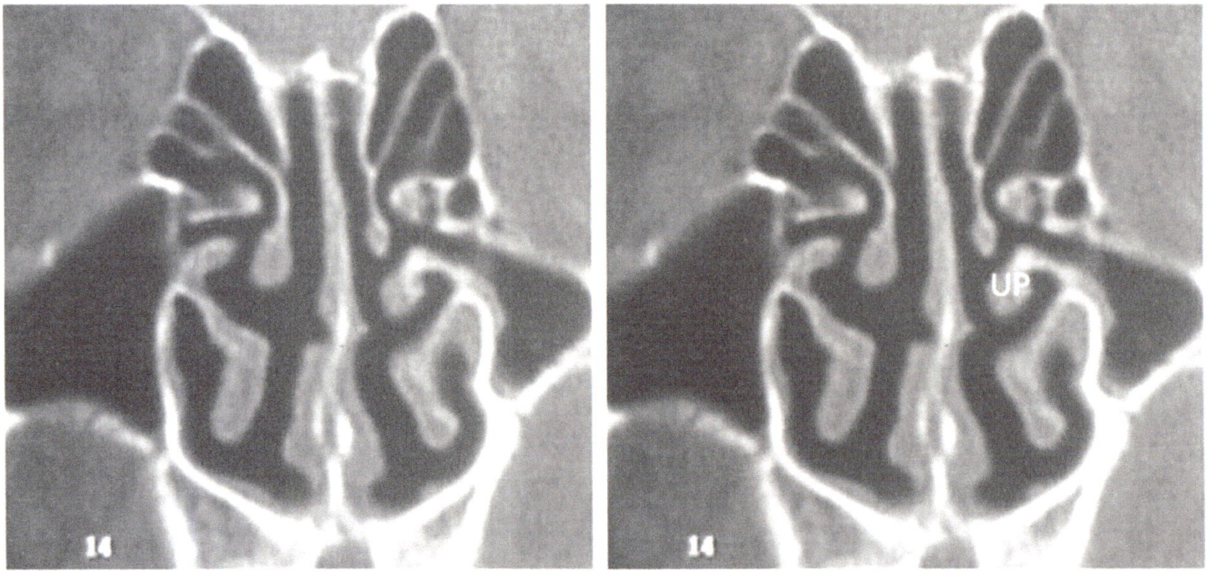

Fig. 4.7. CT showing "rolled" uncinate process on left side.

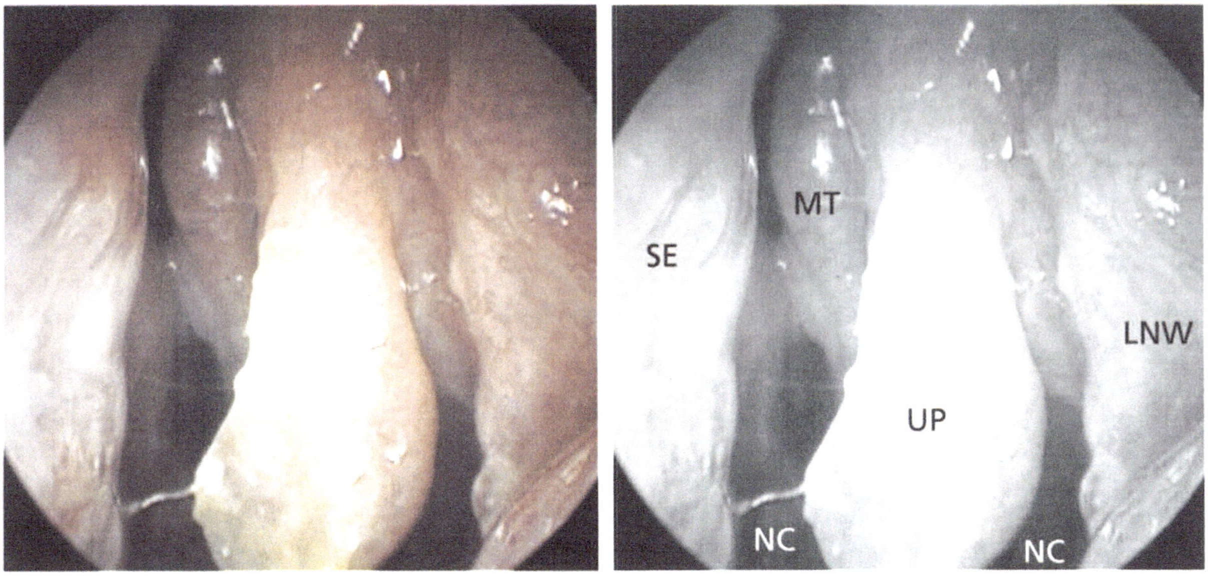

Fig. 4.8. Left side, zero degree lens, corresponding endoscopic finding from previous CT. Uncinate process well anterior to middle turbinate.

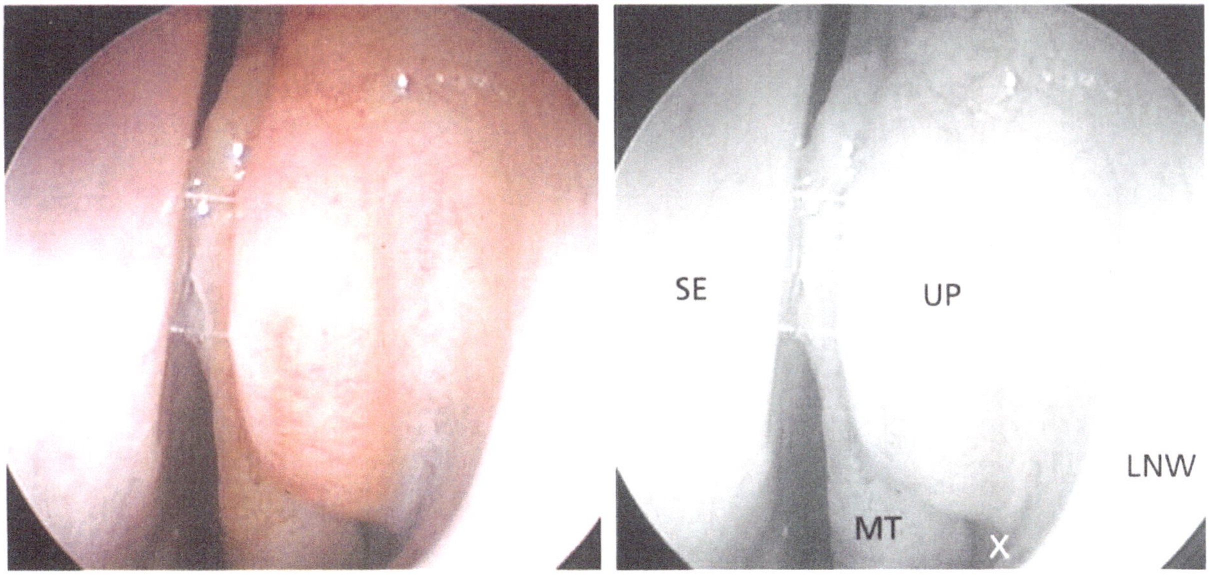

Fig. 4.9. Left side, zero degree lens, anteriorly displaced uncinate process partially covering middle turbinate.

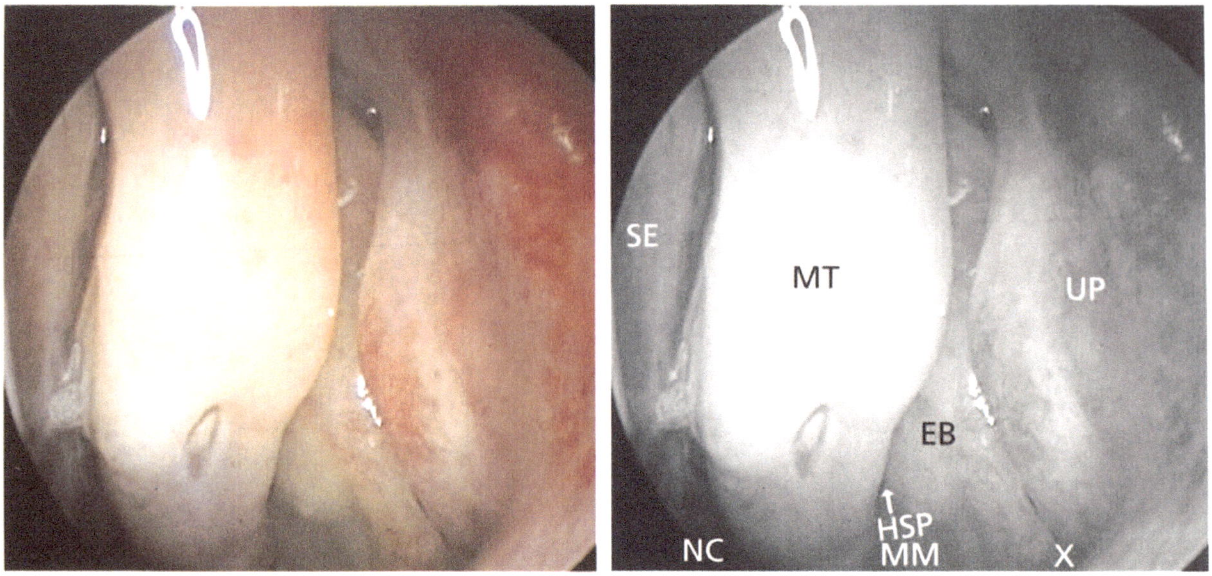

Fig. 4.10. Left side, zero degree lens, large middle meatus, small uncinate process.

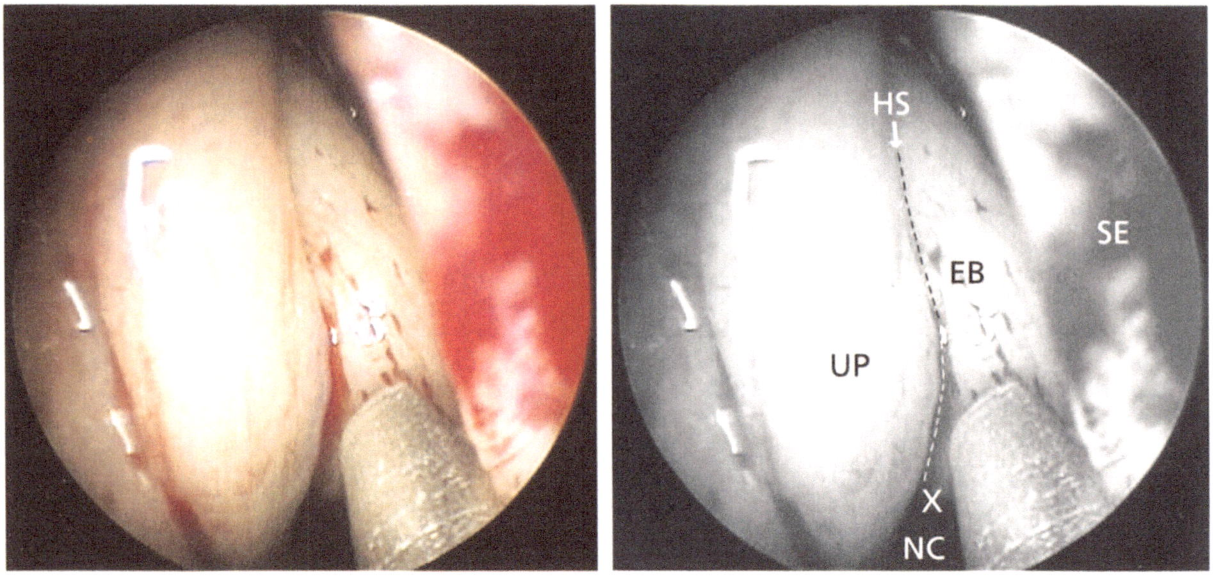

Fig. 4.11. Right side, zero degree lens, uncinate process and ethmoidal bulla with polypoid change.

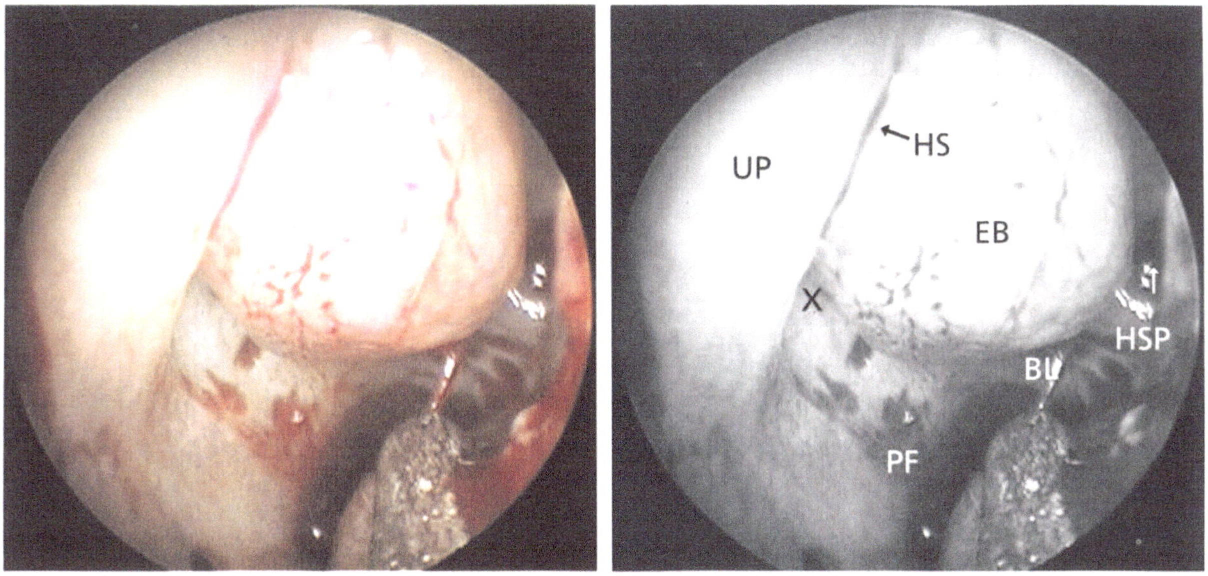

Fig. 4.12. Right side, zero degree lens, same patient showing exit of infundibulum, exposed posterior fontanelle and relationship of uncinate process to ethmoidal bulla.

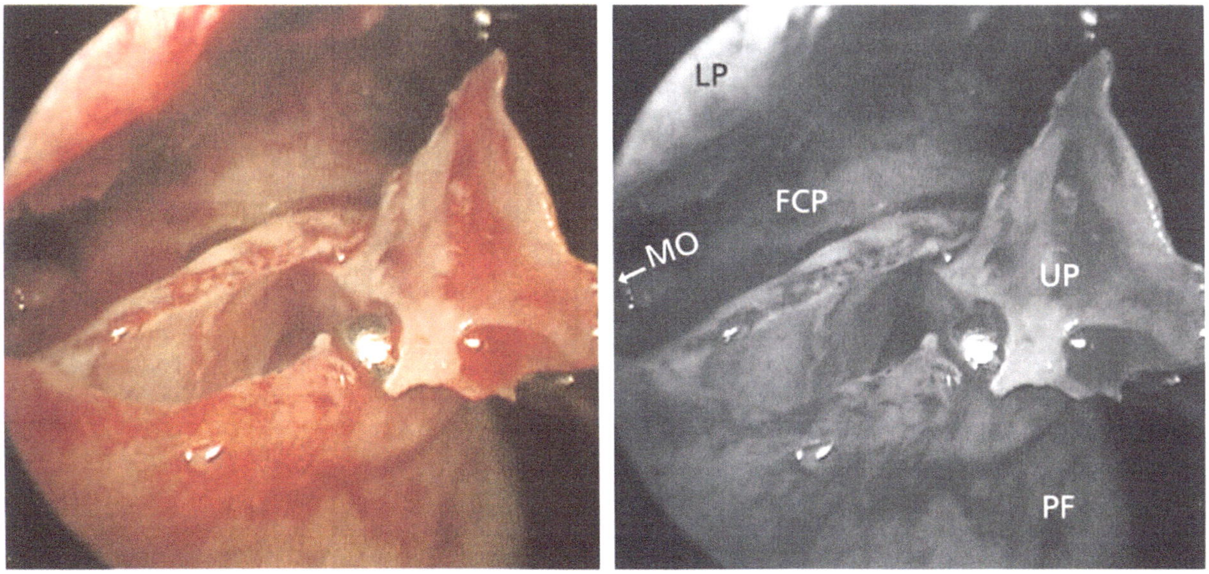

Fig. 4.13. Right side, 30 degree lens, tail of uncinate being removed.

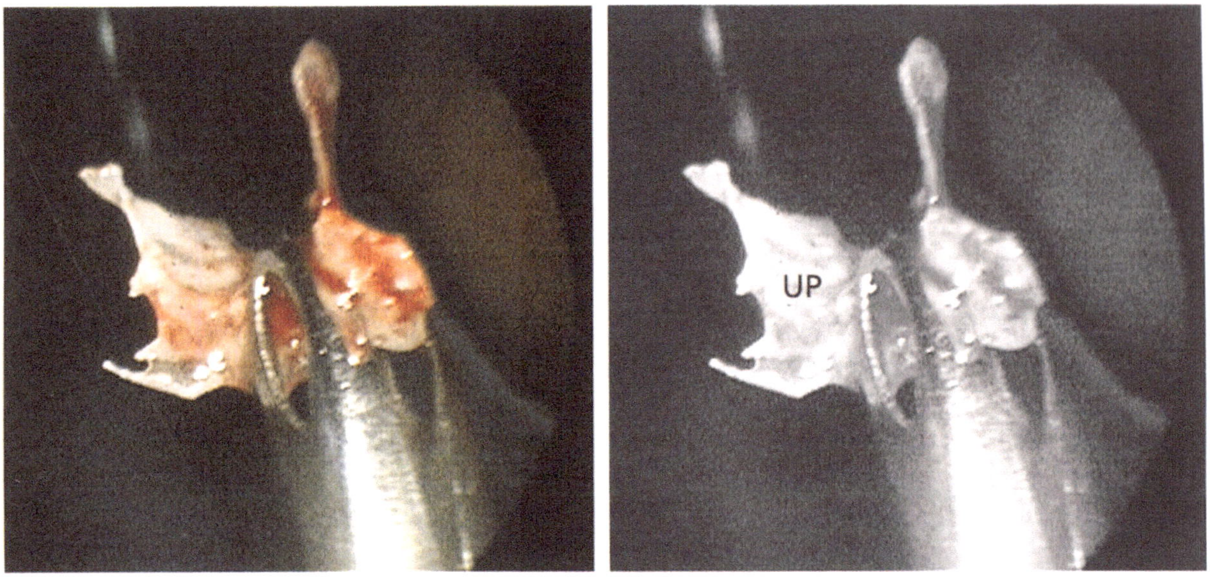

Fig. 4.14. Tail of uncinate.

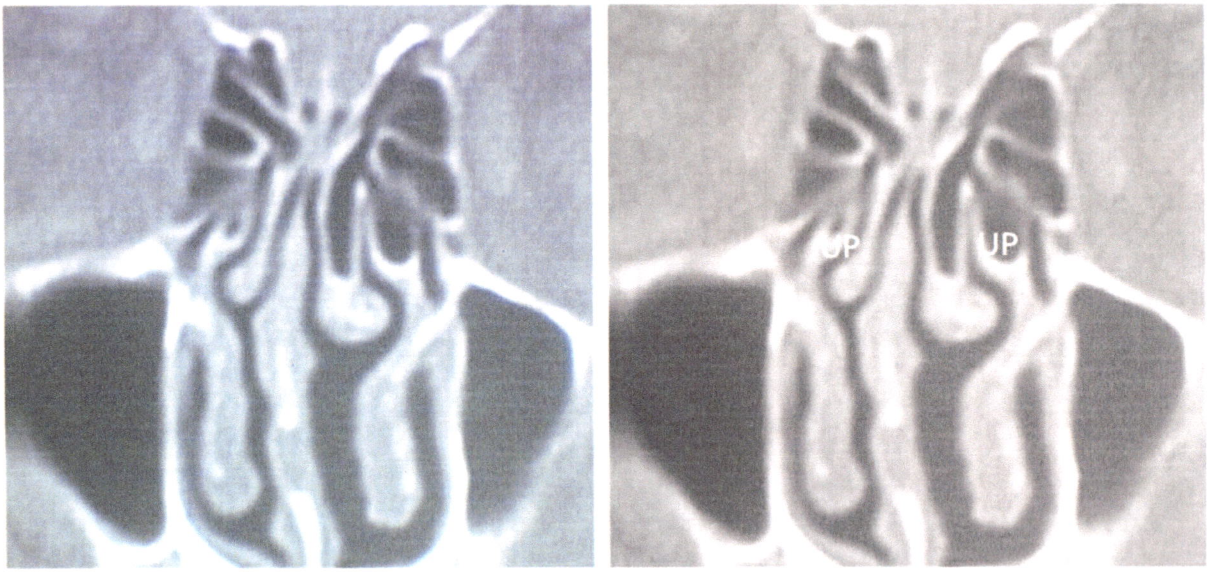

Fig. 4.15. CT, large air cell and uncinate process, left; smaller air cell on right.

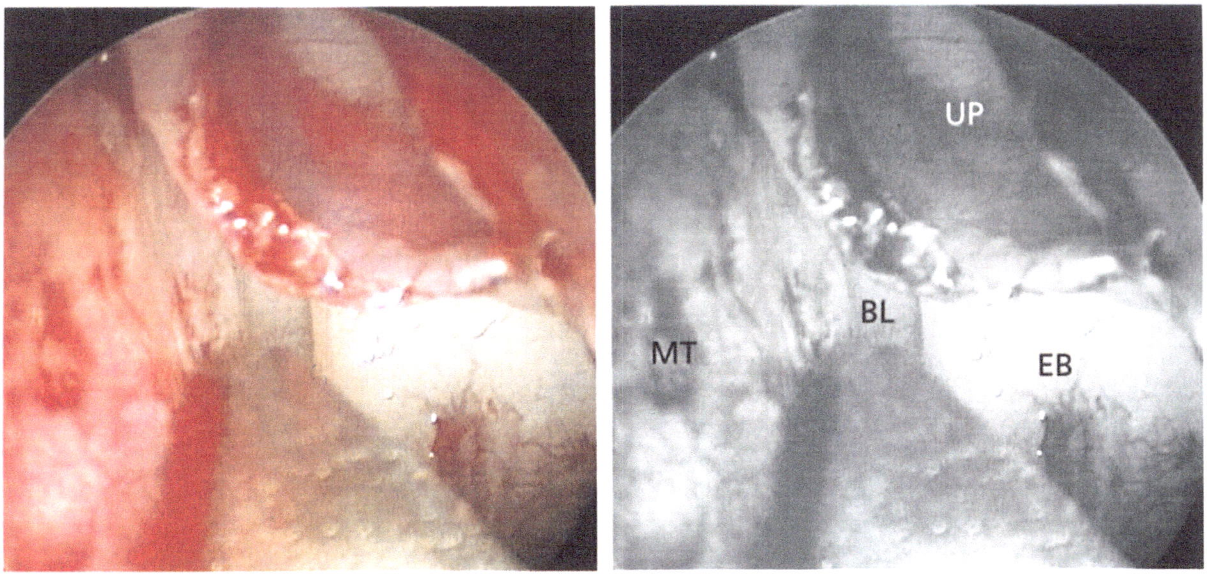

Fig. 4.16. Left side, zero degree lens, ethmoidal bulla, basal lamella, and air cell of uncinate process.

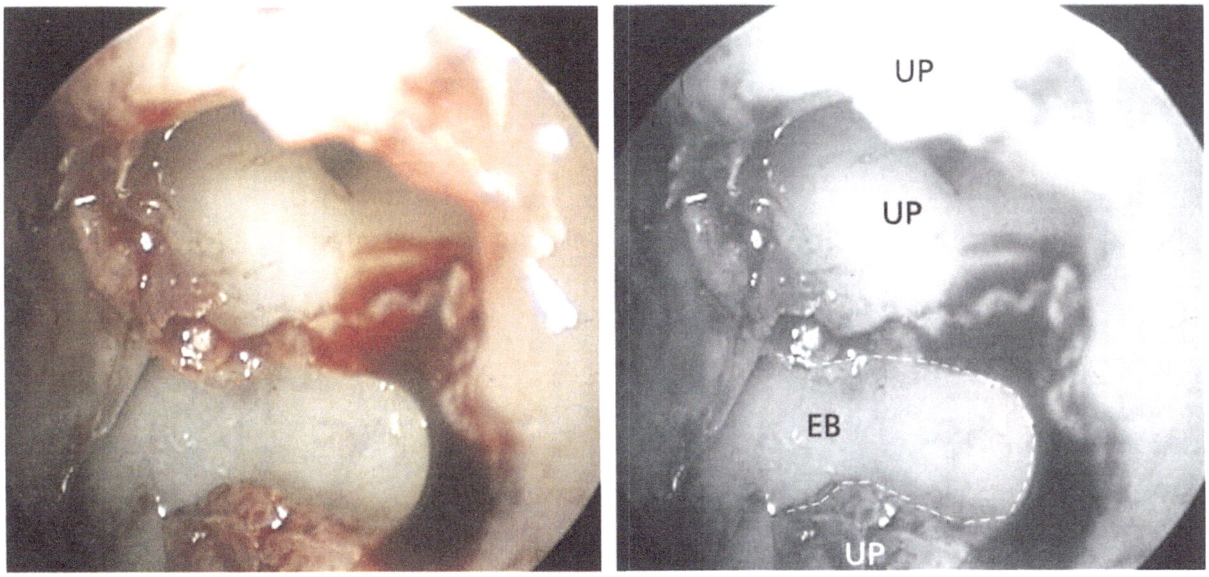

Fig. 4.17. Different patient, air cell of uncinate process and infundibulum, face of ethmoidal bulla.

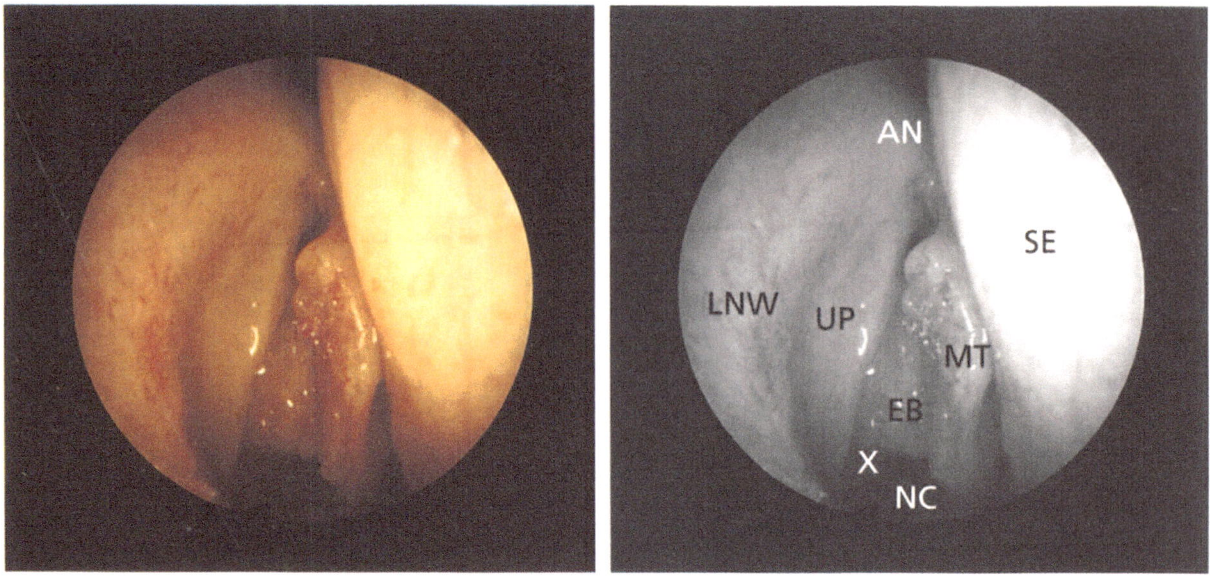

Fig. 4.18. Right side, zero degree lens, middle turbinate partially removed; relationship of uncinate process to bulla is typical.

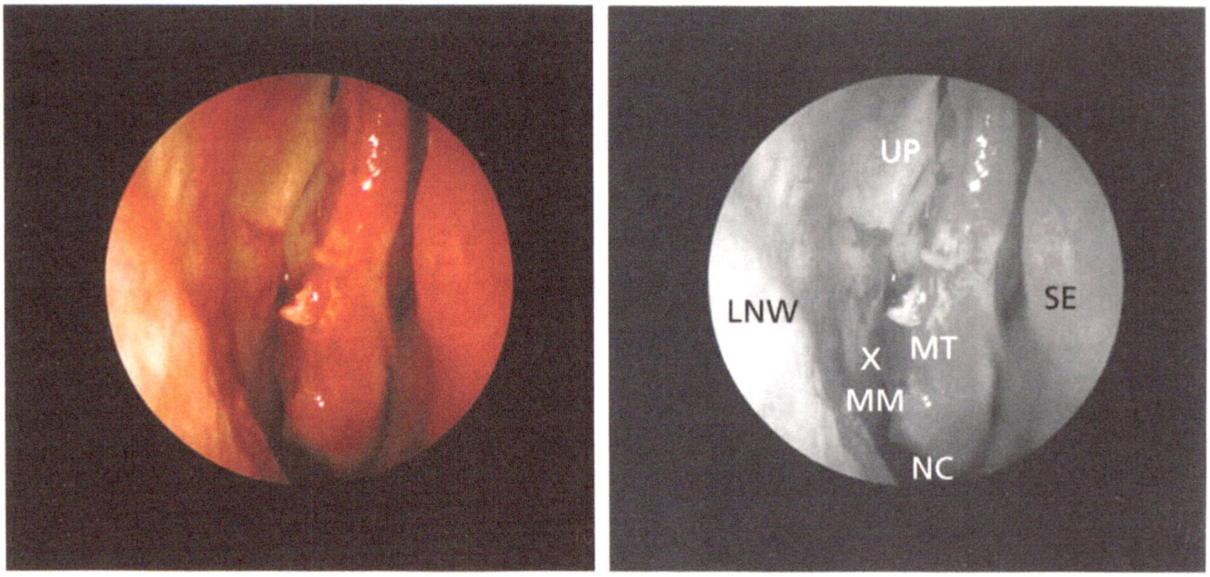

Fig. 4.19. Right side, zero degree lens, middle turbinate partially removed, uncinate process and bulla visible.

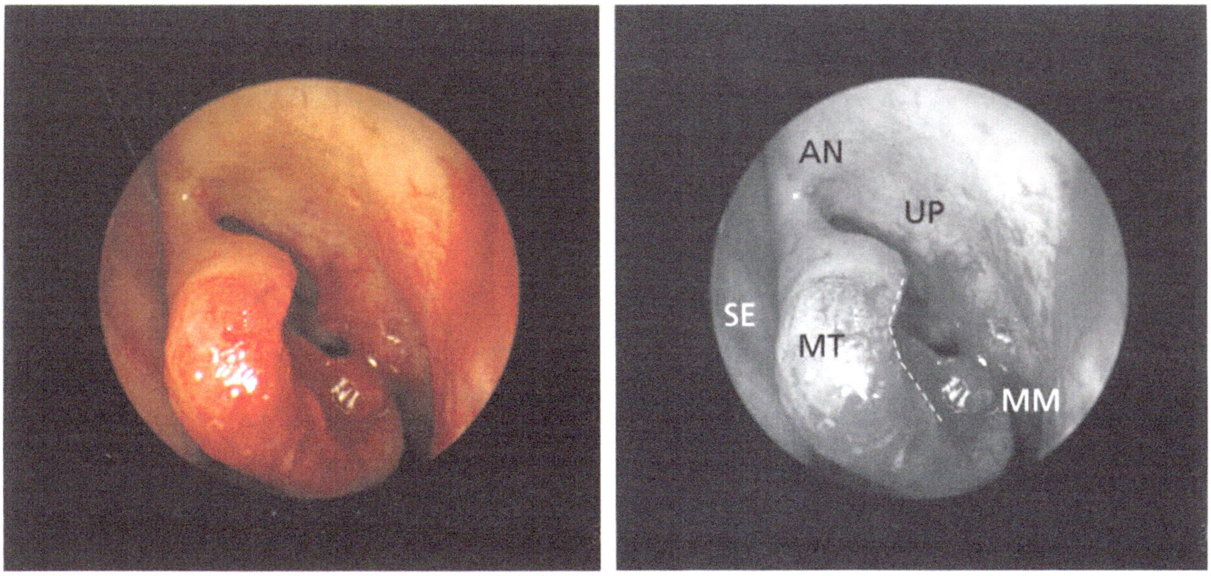

Fig. 4.20. Left side, zero degree lens, partial resection of concha bulla exposing uncinate process.

5. The hiatus semilunaris and infundibulum

The infundibulum is an air-containing space that is bounded by the uncinate process medially, the lamina papyracea laterally, and the ethmoidal bulla posteriorly. The hiatus semilunaris is two-dimensional, defining the extreme medial opening of the infundibulum. Secretions collected by the infundibulum are carried to its exit located at the most inferior and medial aspect of the hiatus semilunaris. The exit is almost always visible at endoscopy between the most inferior-posterior portion of the uncinate process and the ethmoidal bulla.

A constant contributor of secretions to the infundibulum is the maxillary sinus. A Haller cell, when present, may drain either into the infundibulum or into the maxillary sinus. The drainage for the frontal and anterior ethmoid cells is variable with their contributions. The dimensions of the infundibulum, particularly with respect to the approximation of the mucosal surfaces, are dictated by the size and configuration of the ethmoidal bulla, middle turbinate, and the uncinate process. The latter structure varies greatly with respect to its approximation to the lamina papyracea and face of the ethmoidal bulla.

Without the utilization of a 70-degree endoscope, the presence of an accessory opening into the infundibulum, or an anteriorly displaced uncinate process, the space itself cannot be visualized unless the uncinate process is partially or totally removed. The anatomy of the inferior portion of the infundibulum, particularly with respect to the maxillary sinus ostium and the final drainage from the infundibulum, has become more apparent with the use of endoscopes and will be emphasized in this chapter.

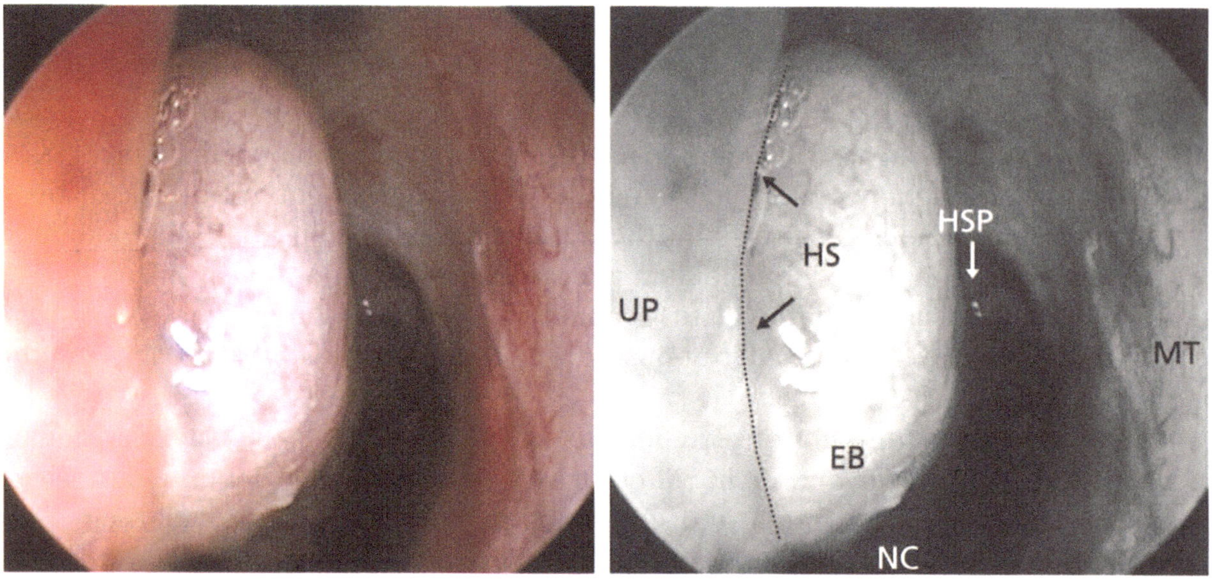

Fig. 5.1. Right side, zero degree lens, bubbles demonstrating the location of the hiatus semilunaris.

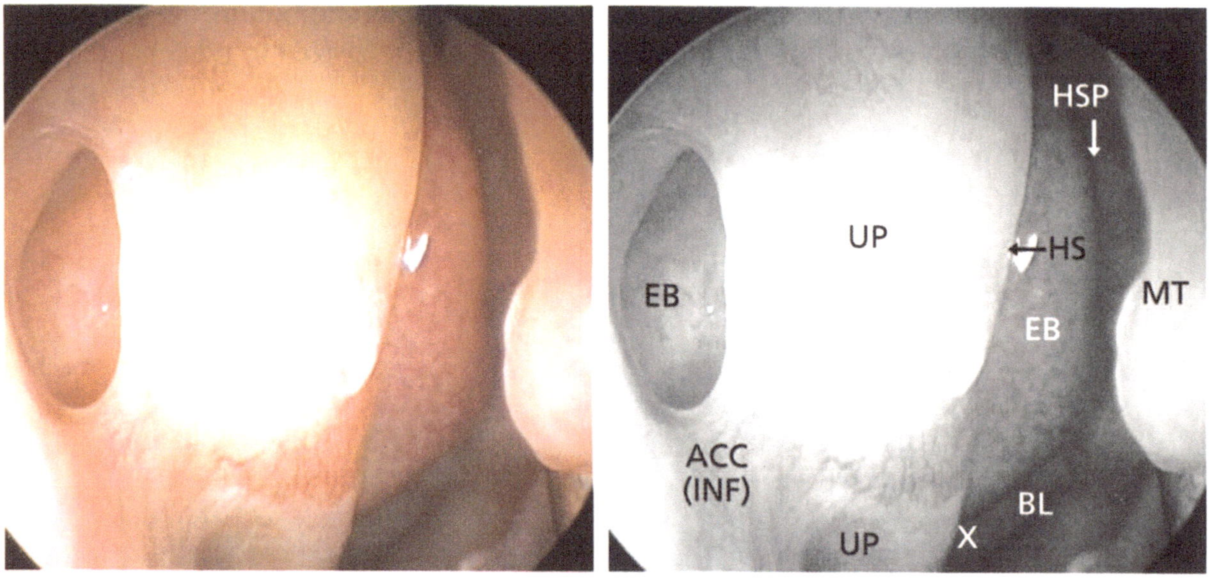

Fig. 5.2. Right side, zero degree lens, accessory opening through uncinate process allows view of ethmoidal bulla in its infundibular portion as well as its middle meatal portion.

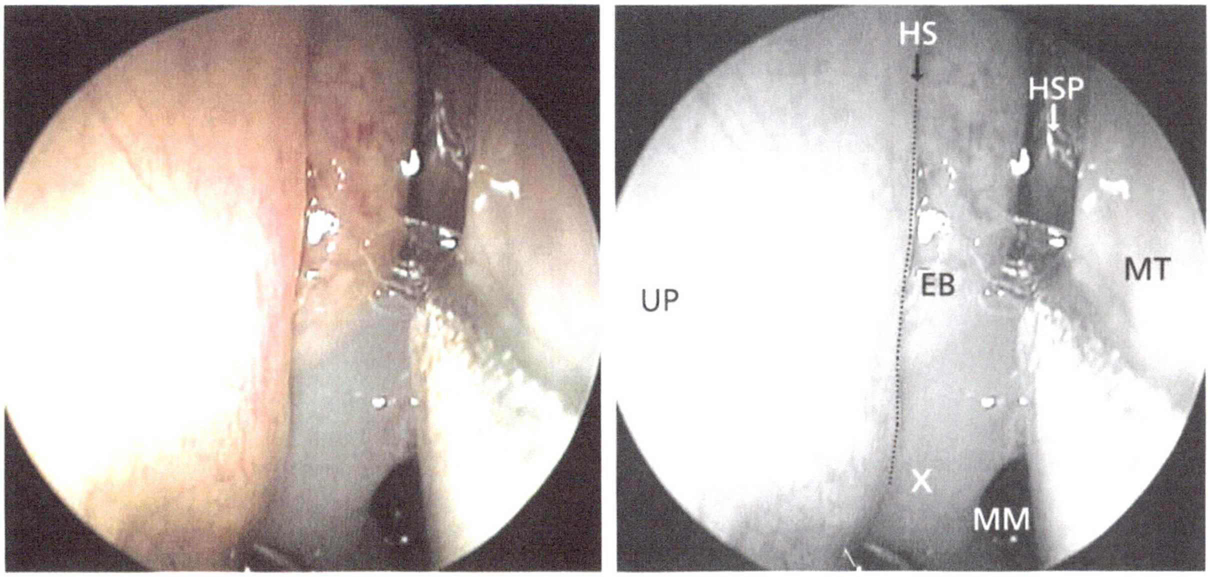

Fig. 5.3. Right side, zero degree lens, secretions from exit of infundibulum.

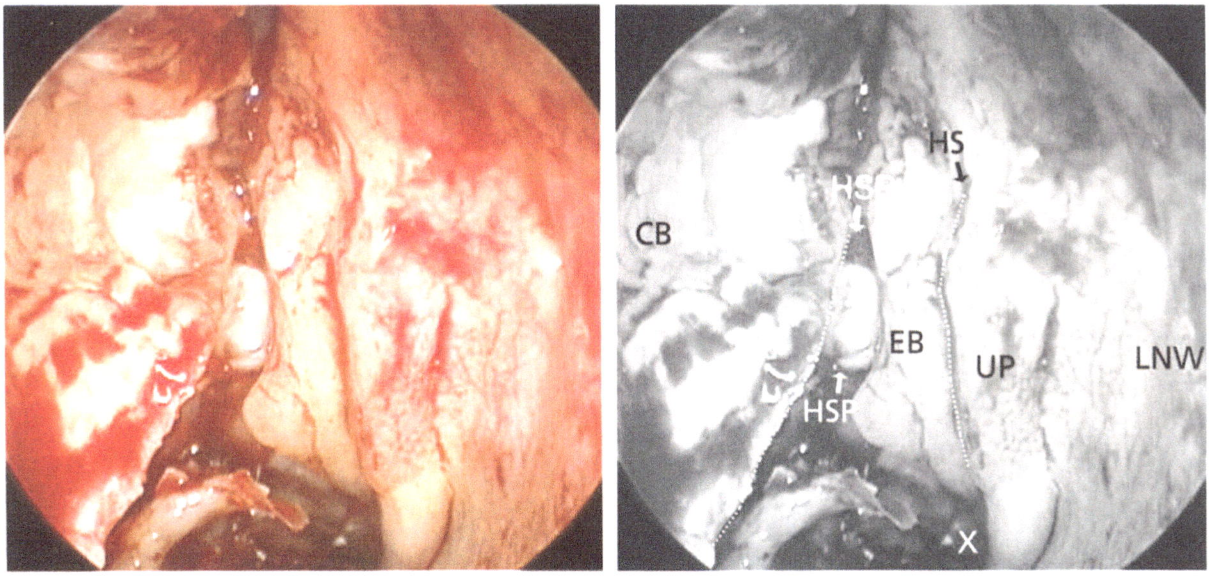

Fig. 5.4. Left side, zero degree lens, partial resection of concha bulla, overview of bulla and uncinate process.

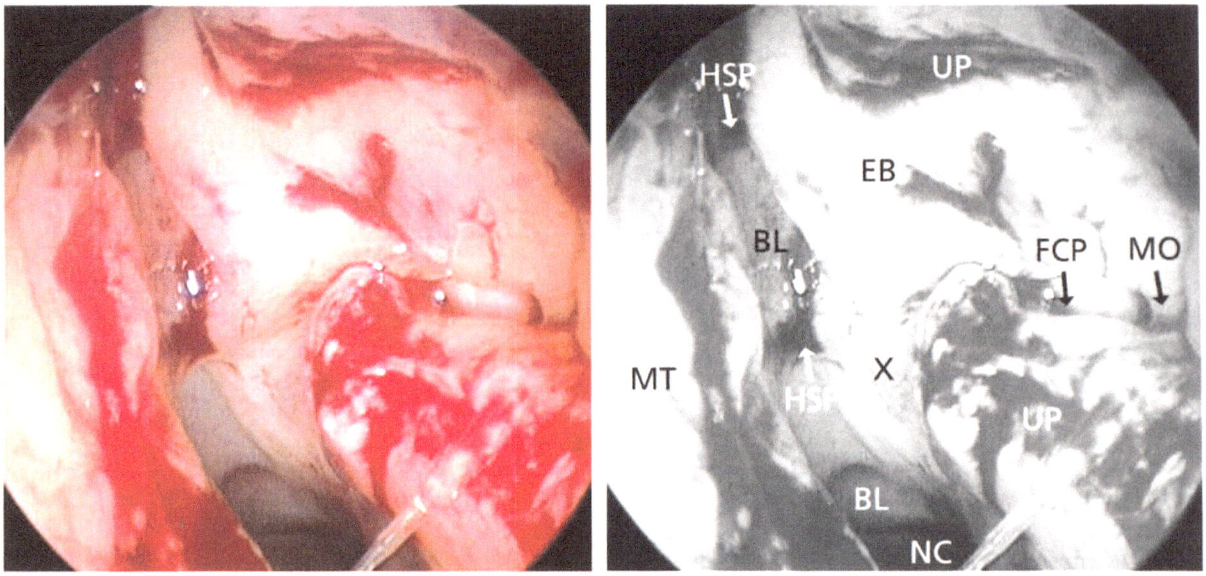

Fig. 5.5. Left side, "window" into inferior portion of infundibulum, secretion from maxillary ostium.

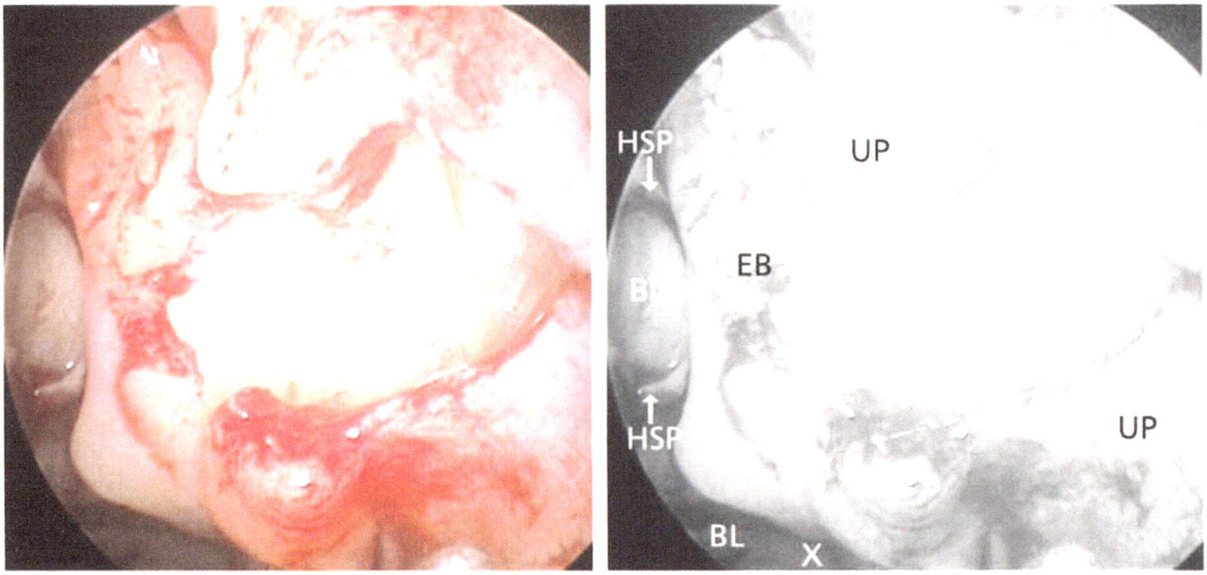

Fig. 5.6. Left side, zero degree lens, similar window showing cut edges of uncinate process and drainage coming to the exit of infundibulum. An extension of this window and retraction of lower cut edge of uncinate process will be required to view the maxillary sinus ostium, final common pathway, and exit of infundibulum.

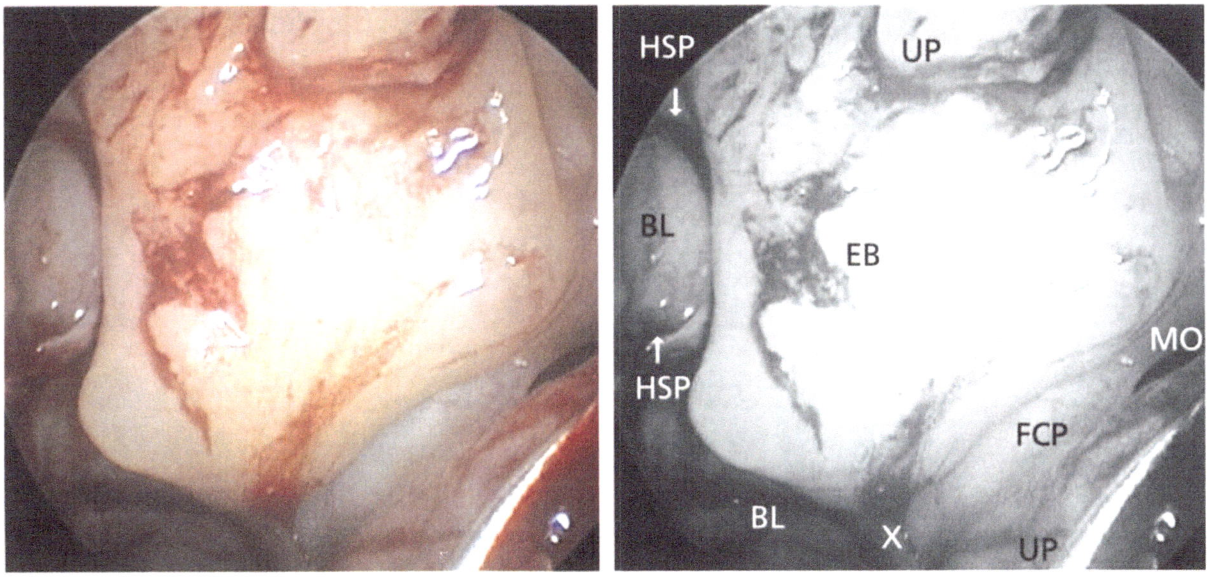

Fig. 5.7. Left side, zero degree lens, after extension of window, relationship of lower uncinate process, and ethmoidal bulla forming final common pathway and exit of infundibulum.

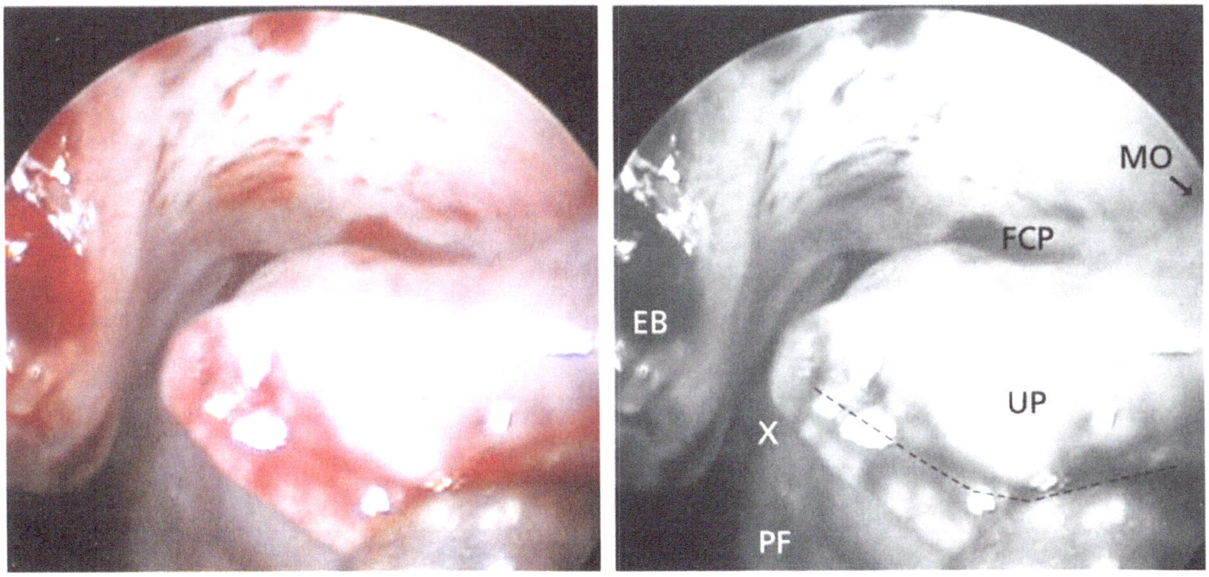

Fig. 5.8. Left side, 30 degree lens, exit of infundibulum (most inferior part of hiatus semilunaris).

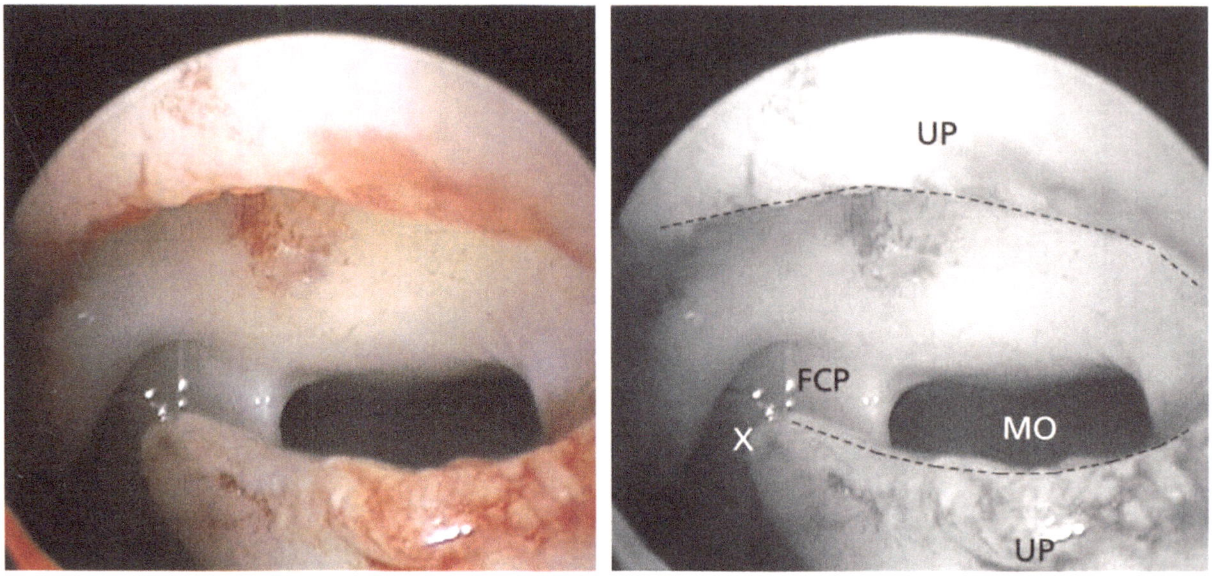

Fig. 5.9. Left side, 30 degree lens, uncinate window with maxillary sinus ostium, final common pathway, and exit of infundibulum.

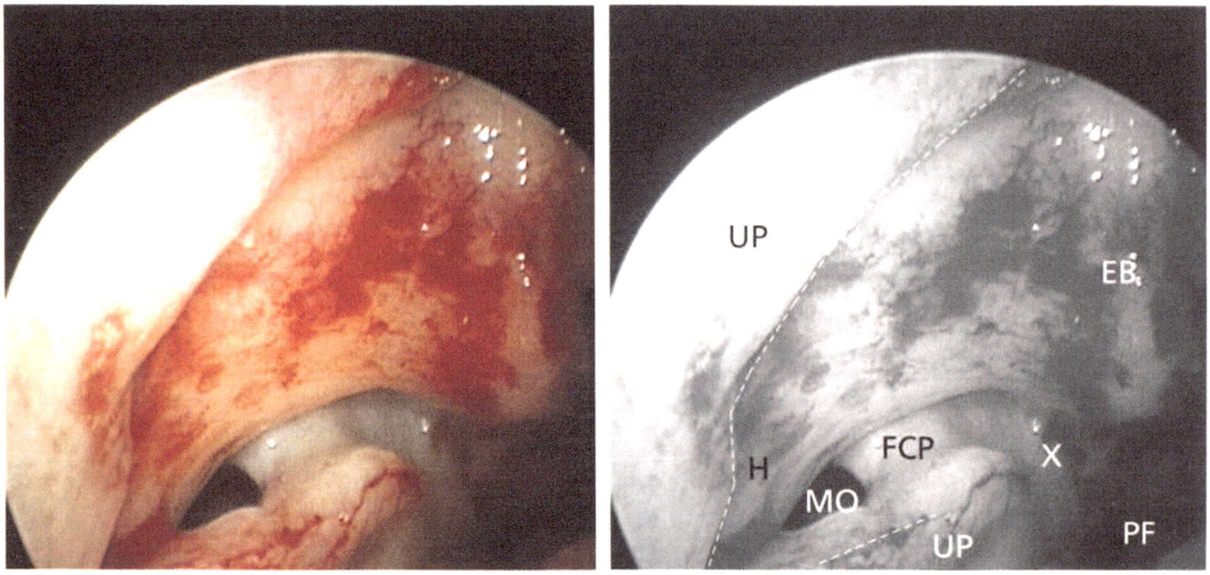

Fig. 5.10. Right side, 30 degree lens, similar view with larger window and more complete removal of inferior portion of uncinate process.

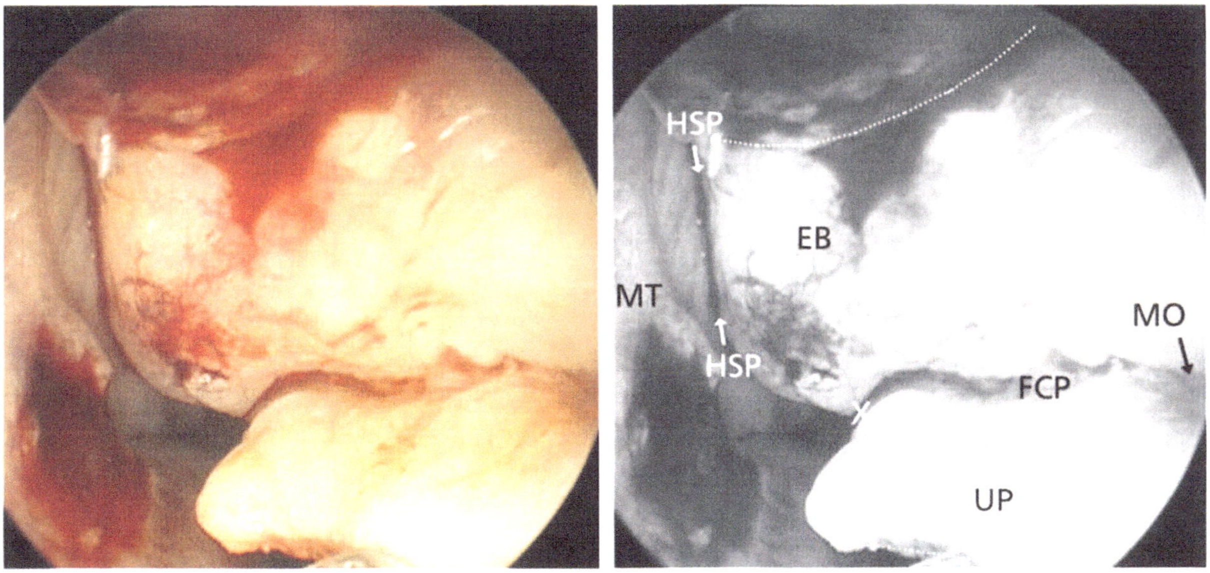

Fig. 5.11. Left side, zero degree lens, a marsupialized lower infundibulum showing face of bulla, final common pathway, and exit. The maxillary sinus ostium is at 3 o'clock. The hiatus semilunaris posterior is seen between the medial wall of the ethmoidal bulla and middle turbinate.

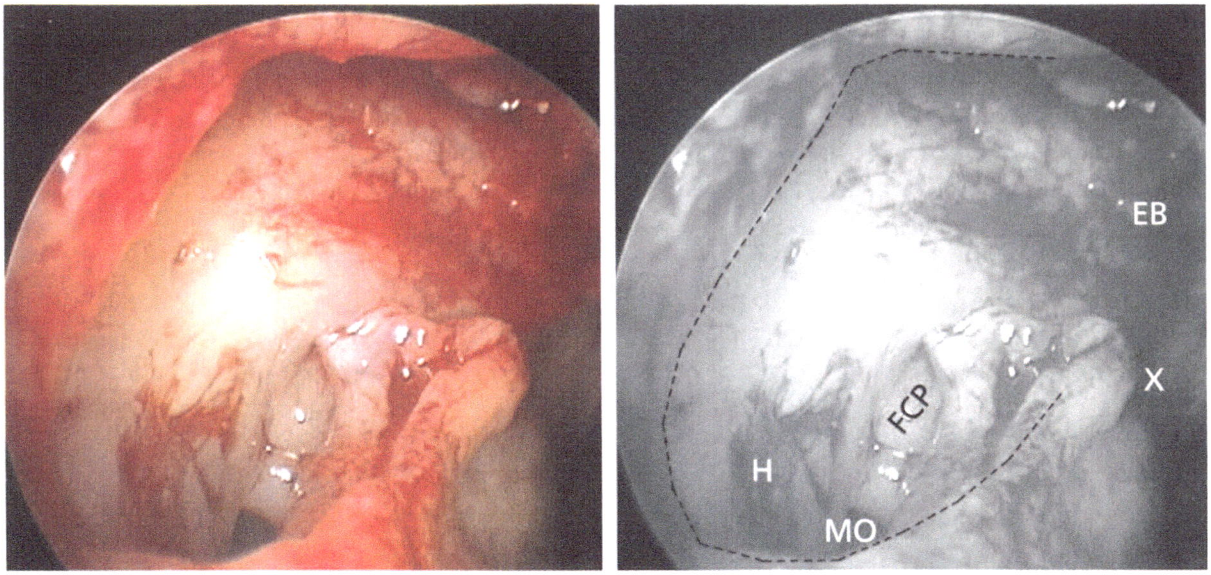

Fig. 5.12. Right side, 30 degree lens, large window showing infundibulum and relationship to maxillary sinus ostium. A Haller cell is present.

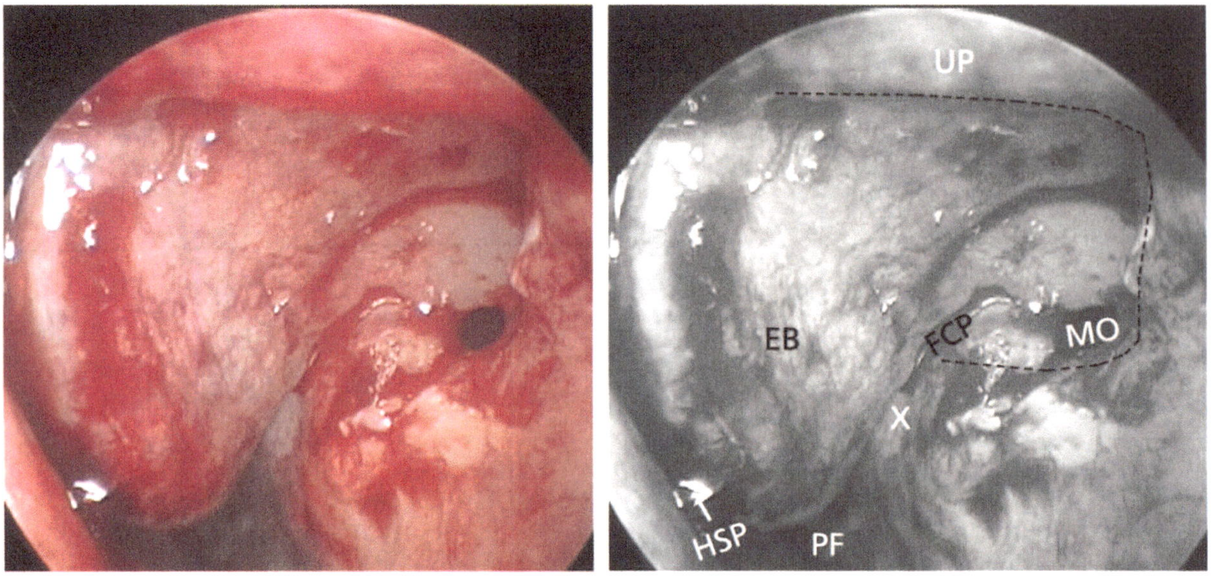

Fig. 5.13. Left side, 30 degree lens, tail of uncinate has been submucosally removed, maxillary ostium empties directly into the nasal cavity; ethmoidal bulla is overhanging posterior fontanelle.

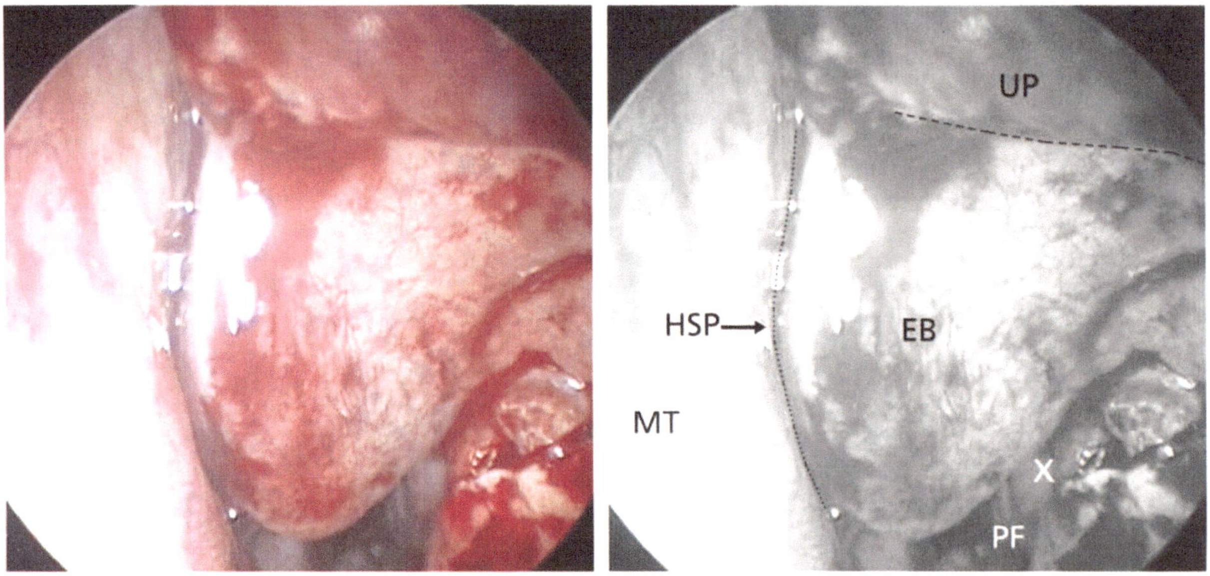

Fig. 5.14. Same specimen as Fig. 5.13, left side, zero degree lens, showing hiatus semilunaris posterior.

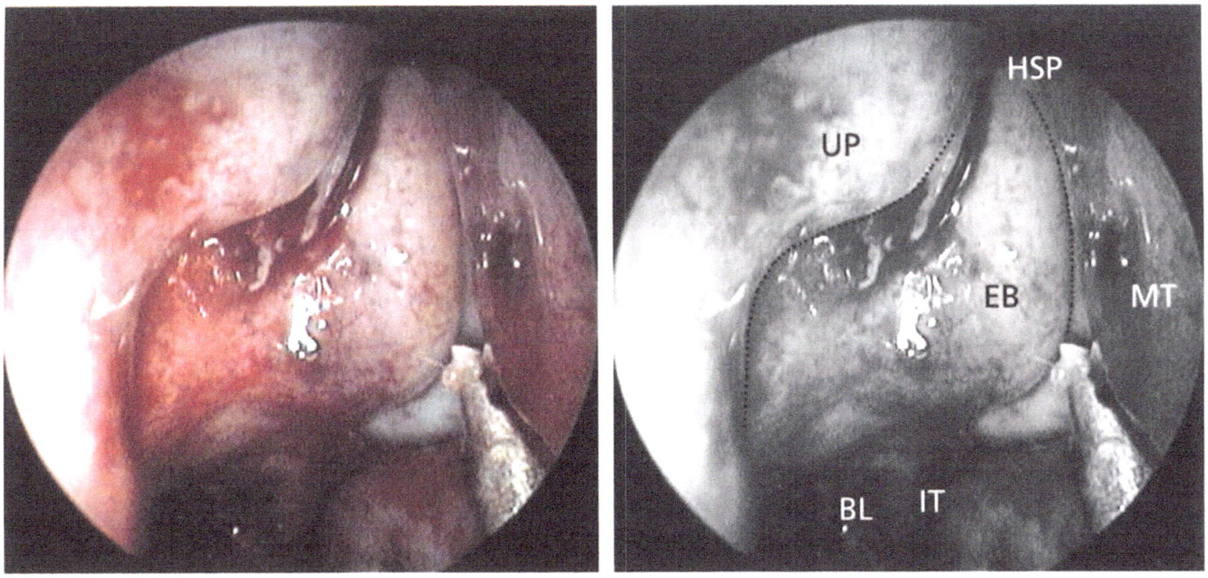

Fig. 5.15. Right side, zero degree lens, hiatus semilunaris posterior, and large bulla.

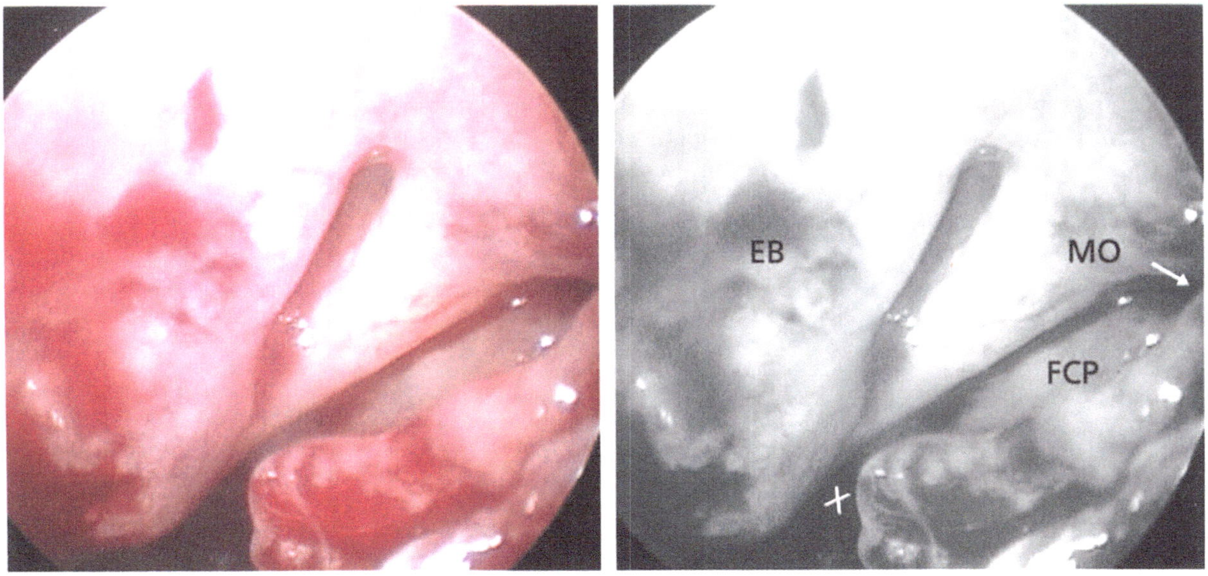

Fig. 5.16. Left side, 30 degree lens, final common pathway, and exit with an ethmoid opening into the infundibulum.

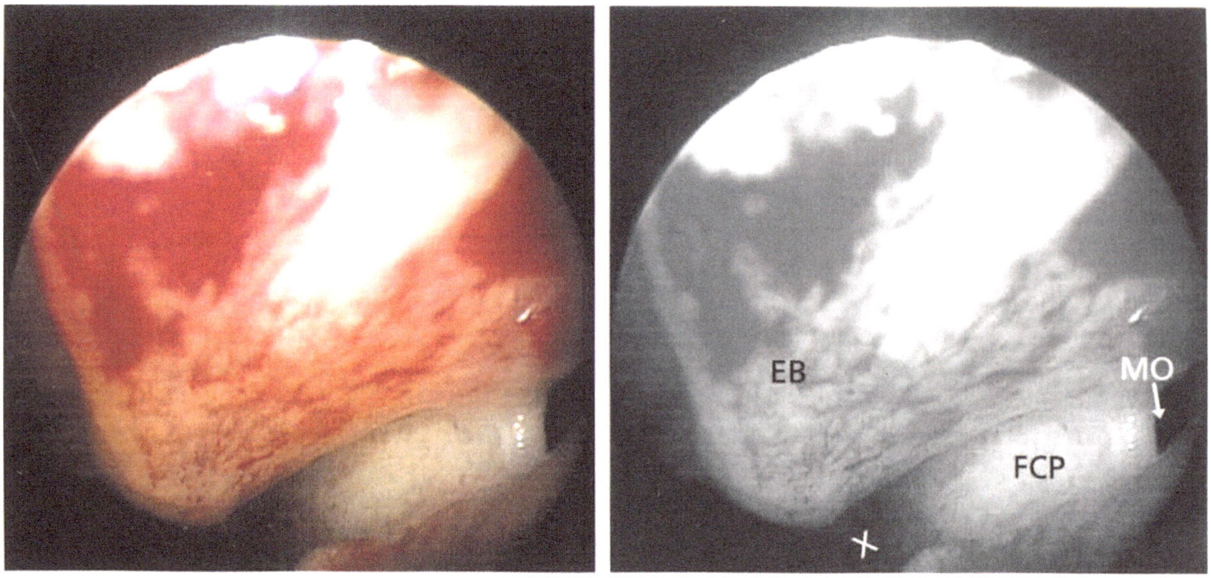

Fig. 5.17. Left side, zero degree lens, relationship of maxillary sinus ostium, final common pathway, exit of infundibulum, and ethmoidal bulla.

6. The ethmoidal bulla

The ethmoidal bulla is the second critical anatomical landmark for the surgeon and the second grand lamella for the anatomist. The degree of pneumatization has direct implications for the dimensions of the infundibulum anteriorly and the hiatus semilunaris posterior medially. The latter is a space of varying dimensions and configuration found at the interface between the medial wall of the ethmoidal bulla and the lateral surface of the middle turbinate.

The bulla may present as a single cell or may contain multiple cells with a variable drainage pattern. The most consistent drainage pattern is from the posterior and medial aspect of the bulla medially into the hiatus semilunaris posterior, anterior to the basal lamella.

The external anterior wall of the bulla serves as the posterior limit of the infundibulum. In many instances, a medial and inferior "teardrop" extension of the bulla may be present. The internal posterior anatomy is variable, sometimes presenting with a posterior bony wall and sometimes having the basal lamella as its posterior limit. The ethmoidal bulla is continuous with the lamina papyracea, which serves as its lateral limit.

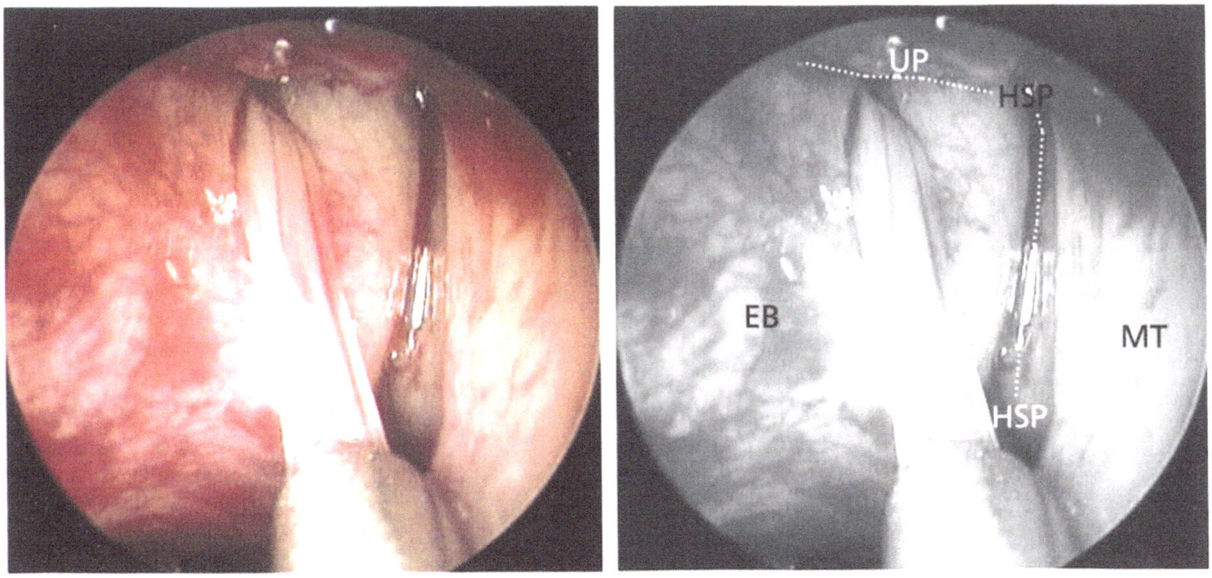

Fig. 6.1. Right side, zero degree lens, secretions from anterior ethmoid cell, hiatus semilunaris posterior to the right. Cut edge of uncinate process at 12 o'clock.

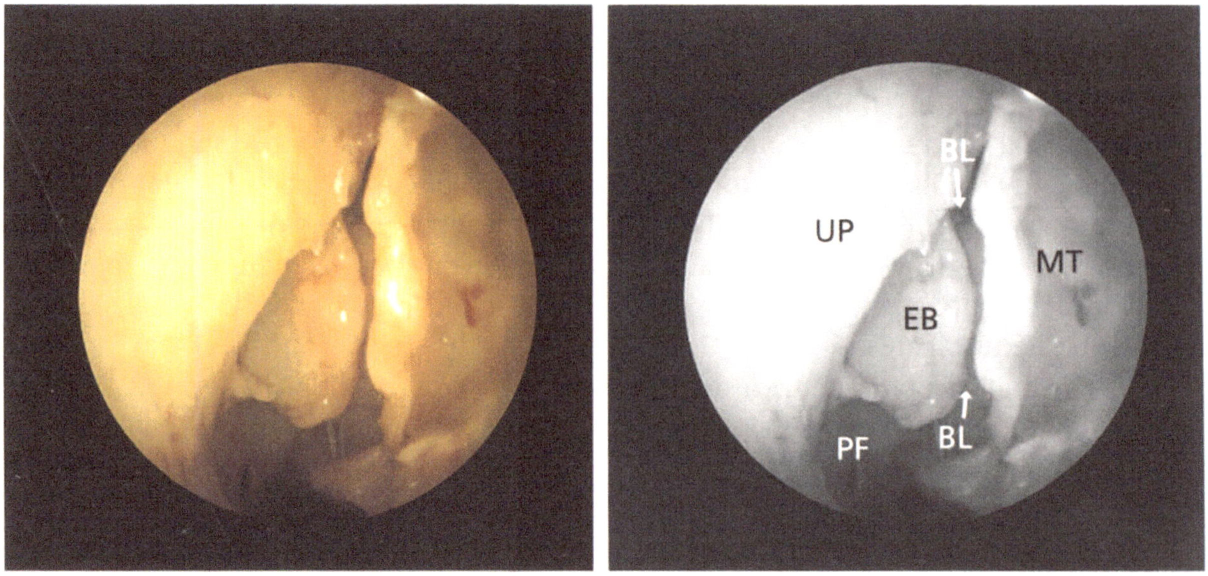

Fig. 6.2. Right side, zero degree lens, ethmoidal bulla seen following partial resection of uncinate process.

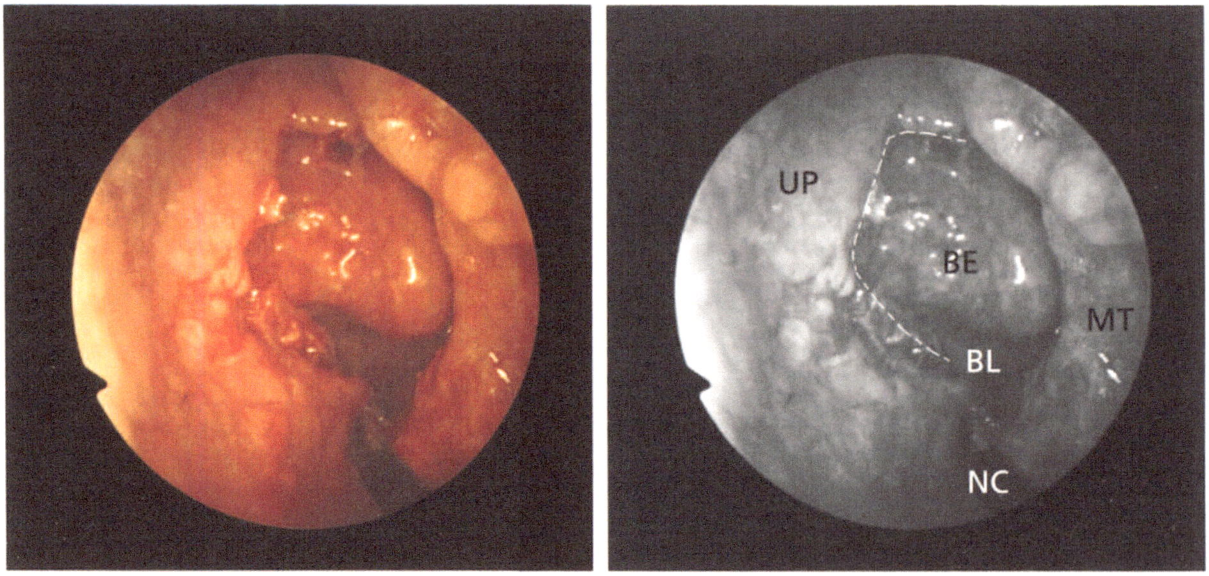

Fig. 6.3. Right side, zero degree lens, ethmoidal bulla; uncinate process has been resected; horizontal portion of basal lamella beneath bulla.

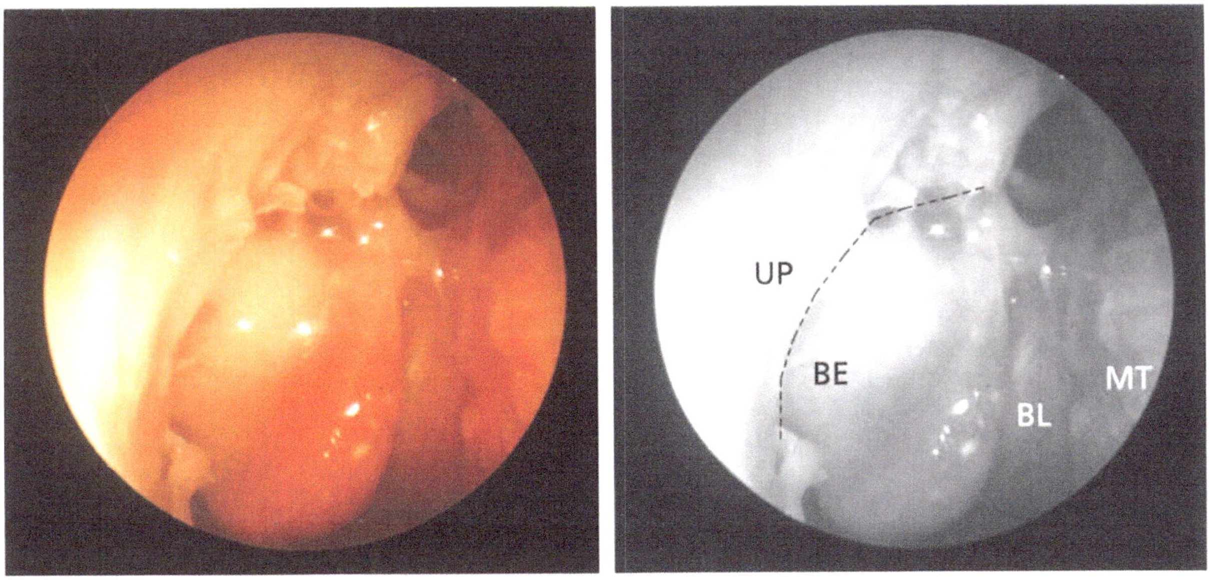

Fig. 6.4. Right side, 30 degree lens, uncinate process resected, ethmoidal bulla, and sinus lateralis.

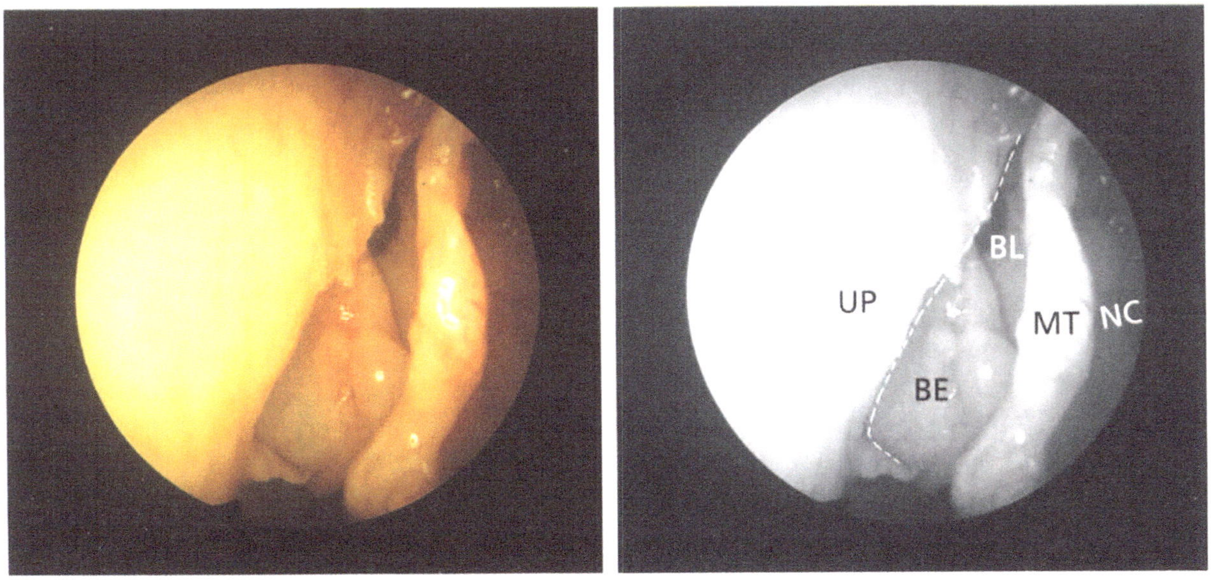

Fig. 6.5. Right side, zero degree lens, ethmoidal bulla following partial uncinectomy.

7. The basal lamella

The basal lamella is the third critical anatomical landmark for the endoscopist and the third grand lamella for the anatomist. The horizontal portion may be visualized within the middle meatus below and posterior to the ethmoidal bulla. The vertical portion is sometimes visualized by simple endoscopy at the posterior limit of the hiatus semilunaris posterior. Most of the basal lamella is inside the cell system of the ethmoid labyrinth posterior to the ethmoidal bulla.

The basal lamella is continuous with both anterior and posterior ethmoid cells and is used to define their separation. The medial extension of the basal lamella becomes the bony part of the middle turbinate itself; laterally it is continuous with the lamina papyracea. The vertical portion may be a wall of bone that swings laterally from the middle turbinate behind the ethmoidal bulla to continue as the lamina papyracea or it may be partially pneumatized. The indentations in the vertical portion of the basal lamella may be found medially, laterally, superiorly, inferiorly, or in some combination thereof.

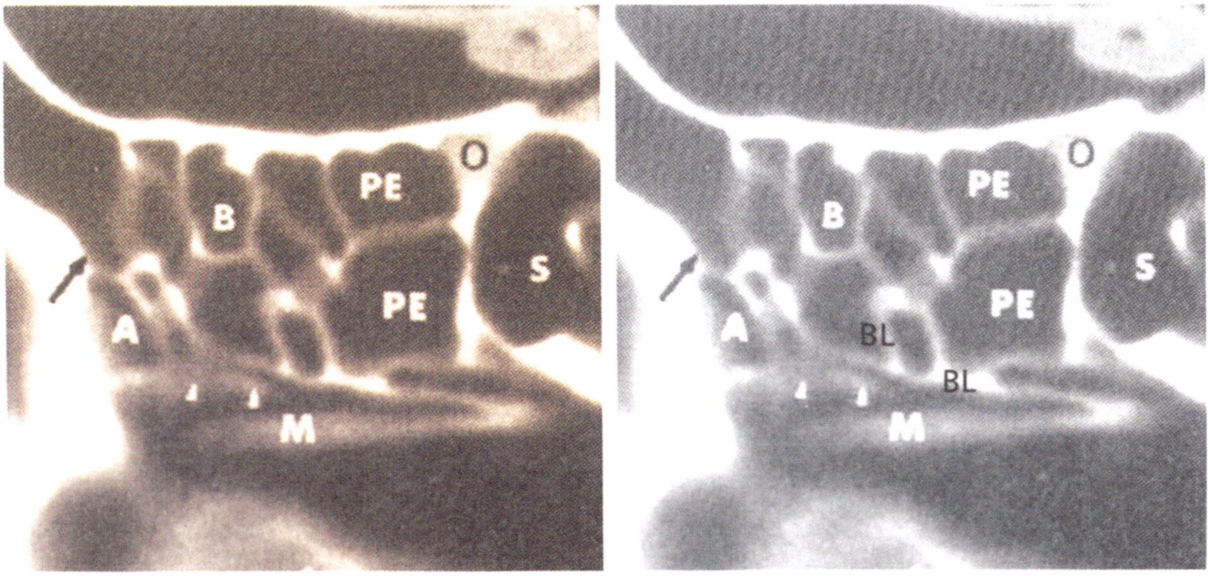

Fig. 7.1. Parasagittal cut through the center of the ethmoid complex showing horizontal and vertical portions of the basal lamella.

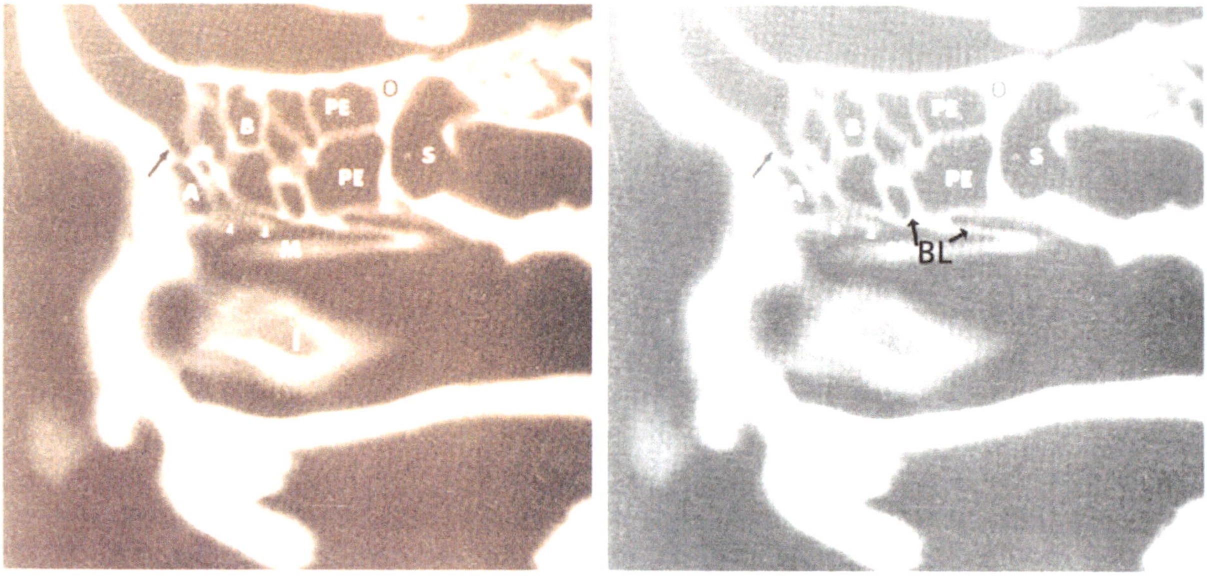

Fig. 7.2. Same as Fig. 7.1 with relationship to inferior turbinate.

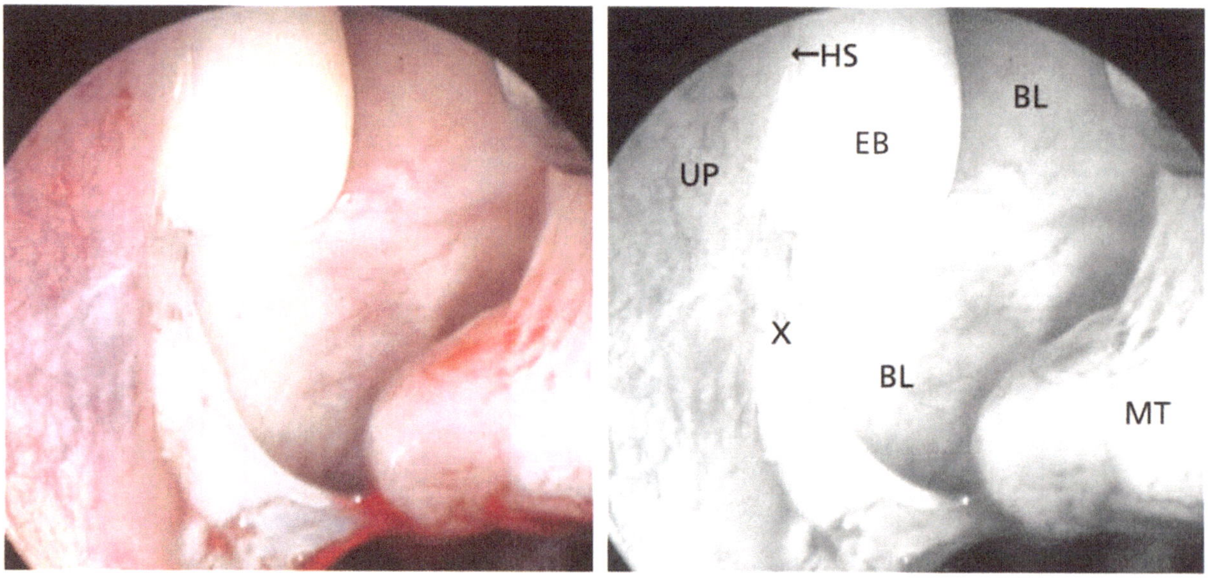

Fig. 7.3. Right side, zero degree lens, secretions from exit of infundibulum. Horizontal and vertical portions of basal lamella visualized by simple endoscopy.

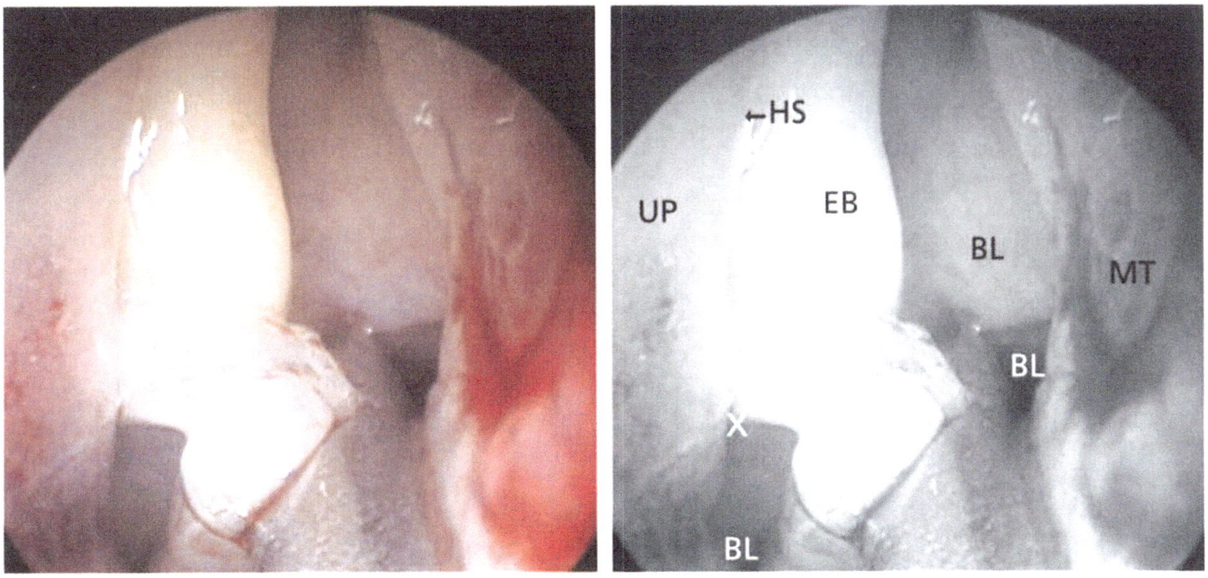

Fig. 7.4. Right side, zero degree lens, same patient, vertical portion of basal lamella visualized lateral to ethmoidal bulla.

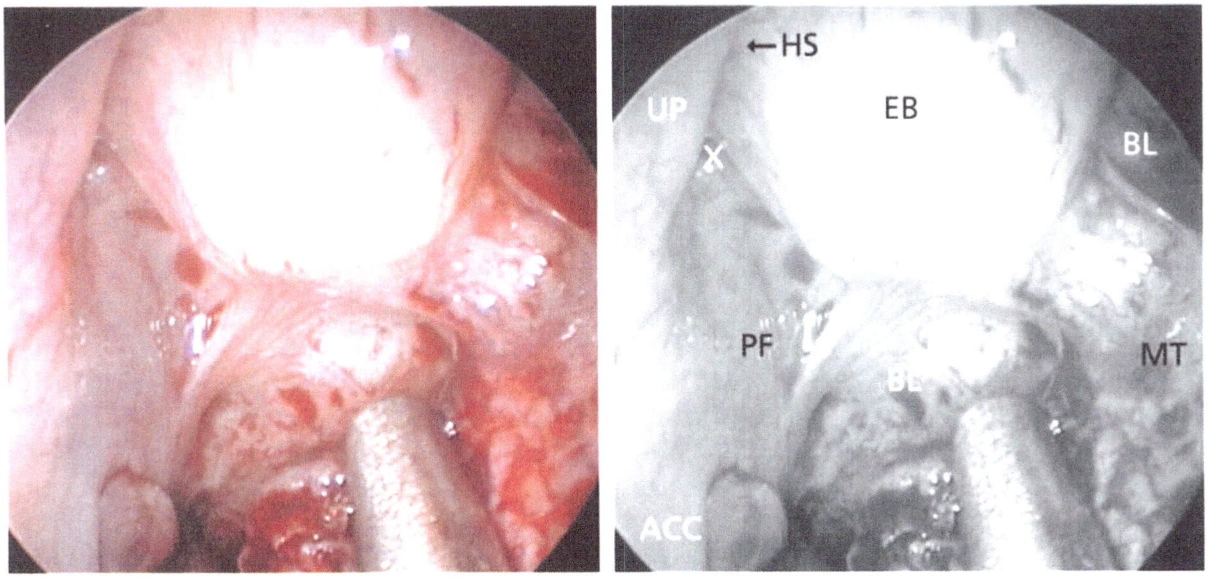

Fig. 7.5. Right side, zero degree lens, horizontal portion of basal lamella and posterior fontanelle, exit of infundibulum at 10 o'clock.

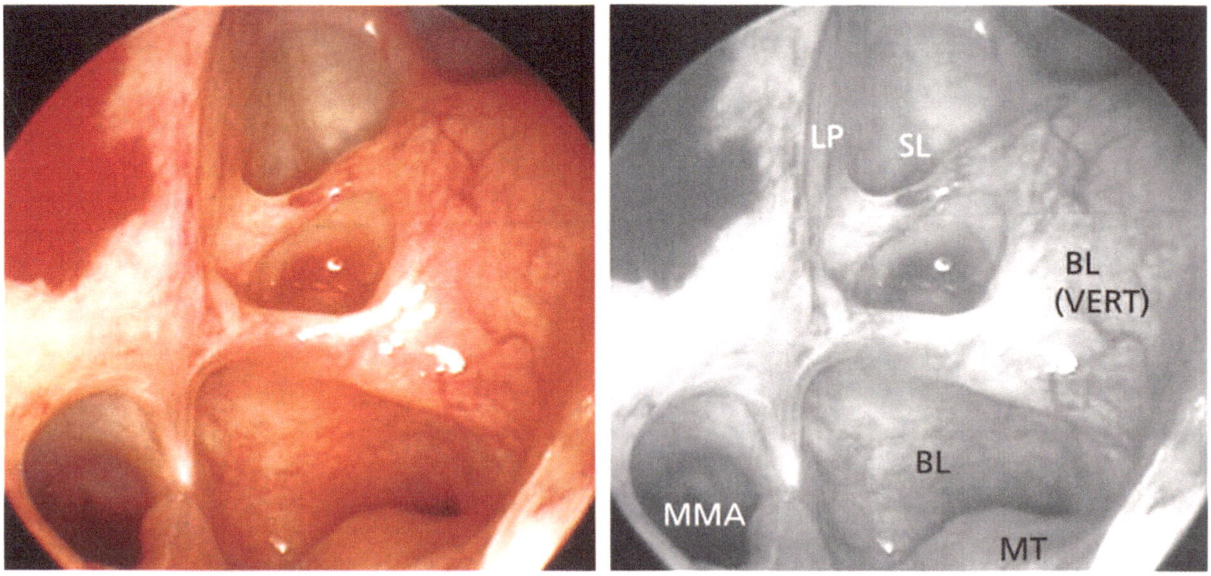

Fig. 7.6. Right side, post-operative, horizontal and vertical portions of basal lamella, pneumatization medially along lamina papyracea.

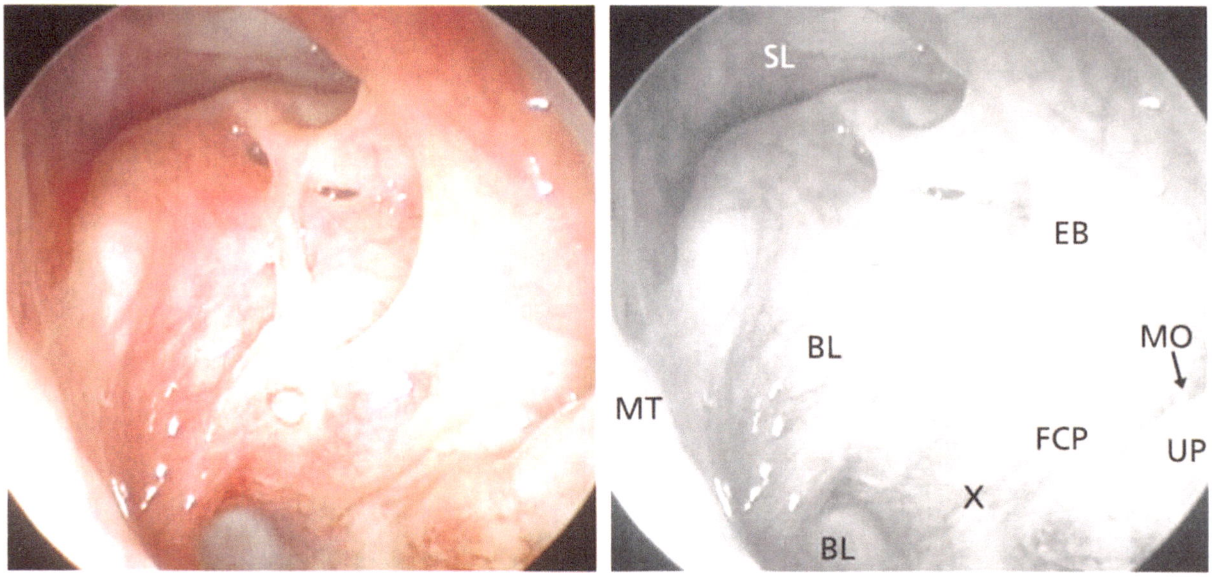

Fig. 7.7. Left side, post-operative, horizontal and vertical portions of basal lamella with pneumatization above.

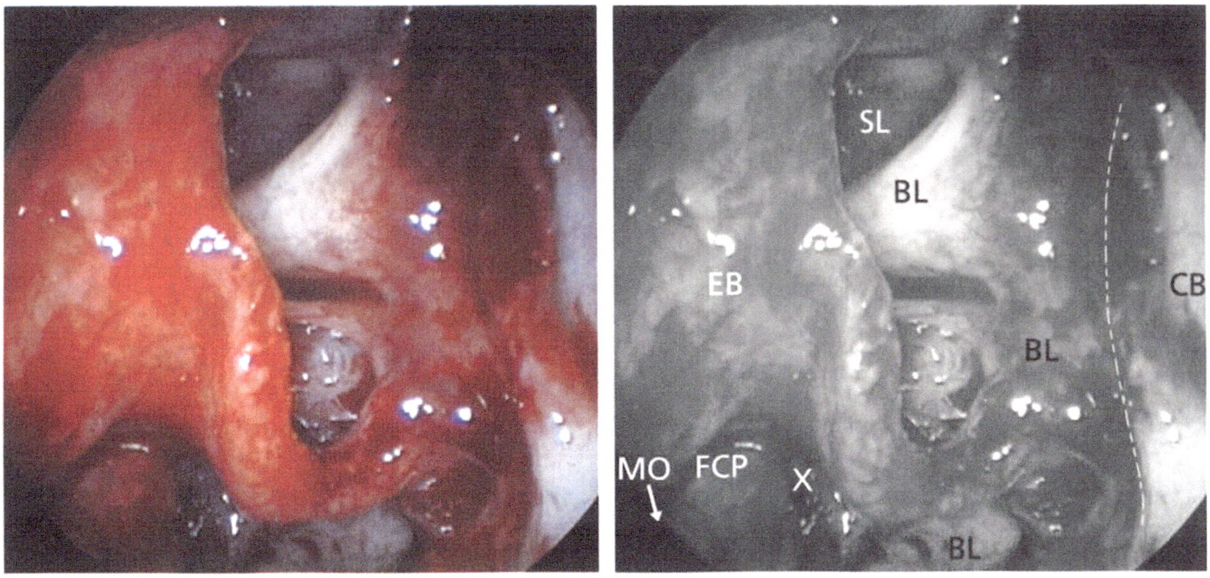

Fig. 7.8. Right side, zero degree lens, horizontal and vertical portions of basal lamella with pneumatization above.

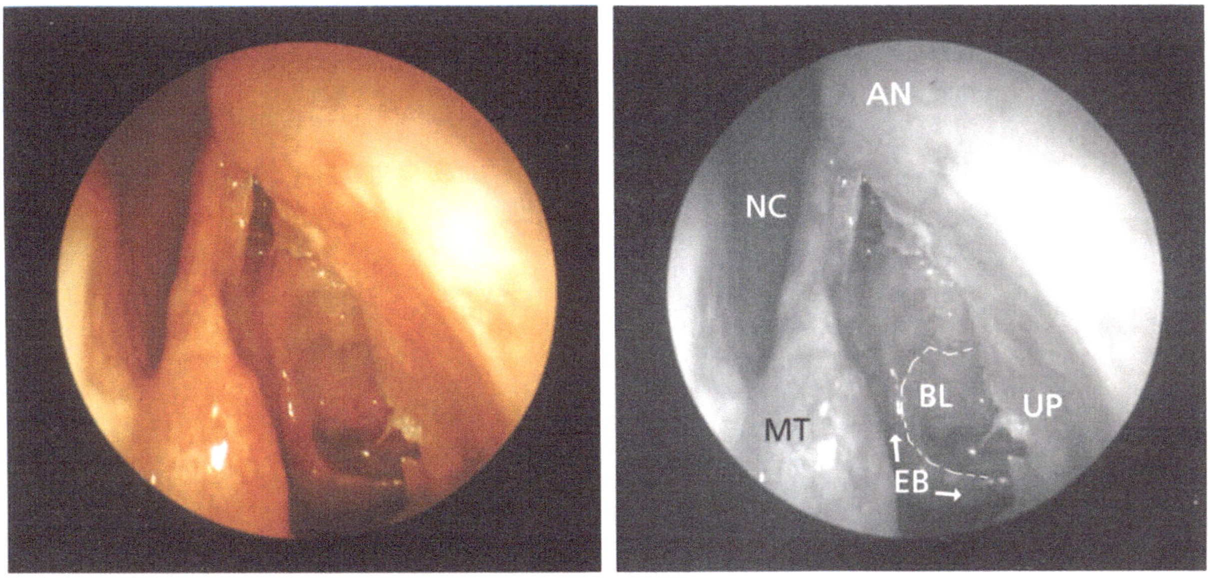

Fig. 7.9. Left side, 30 degree lens, vertical portion of basal lamella seen through open ethmoidal bulla; medial wall of bulla intact.

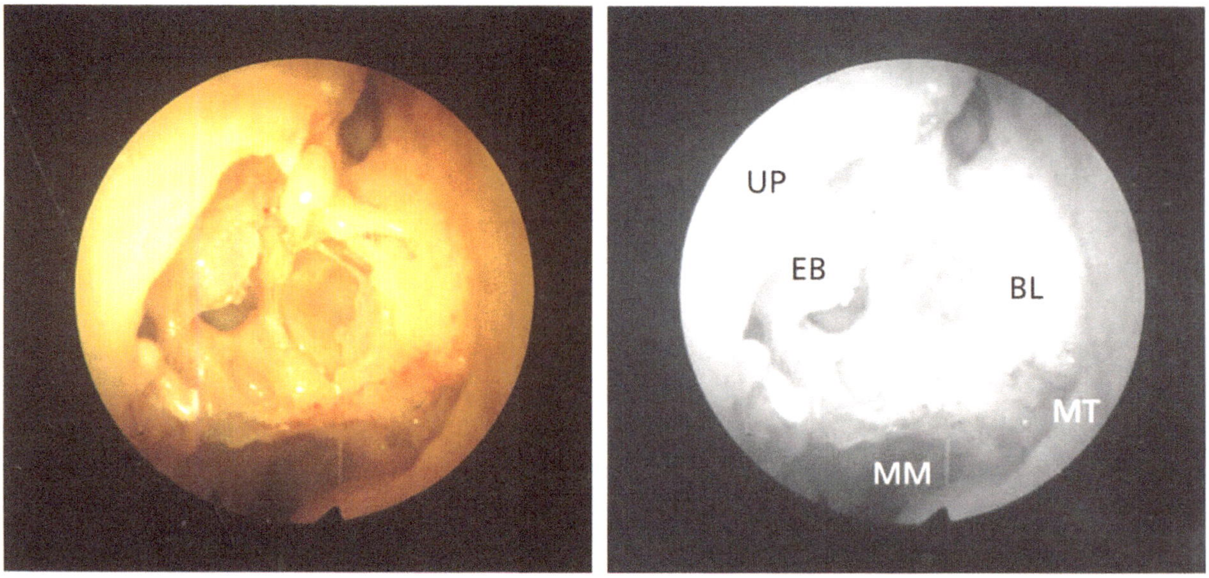

Fig. 7.10. Right side, zero degree lens, partial opening of vertical portion of basal lamella after partial uncinectomy.

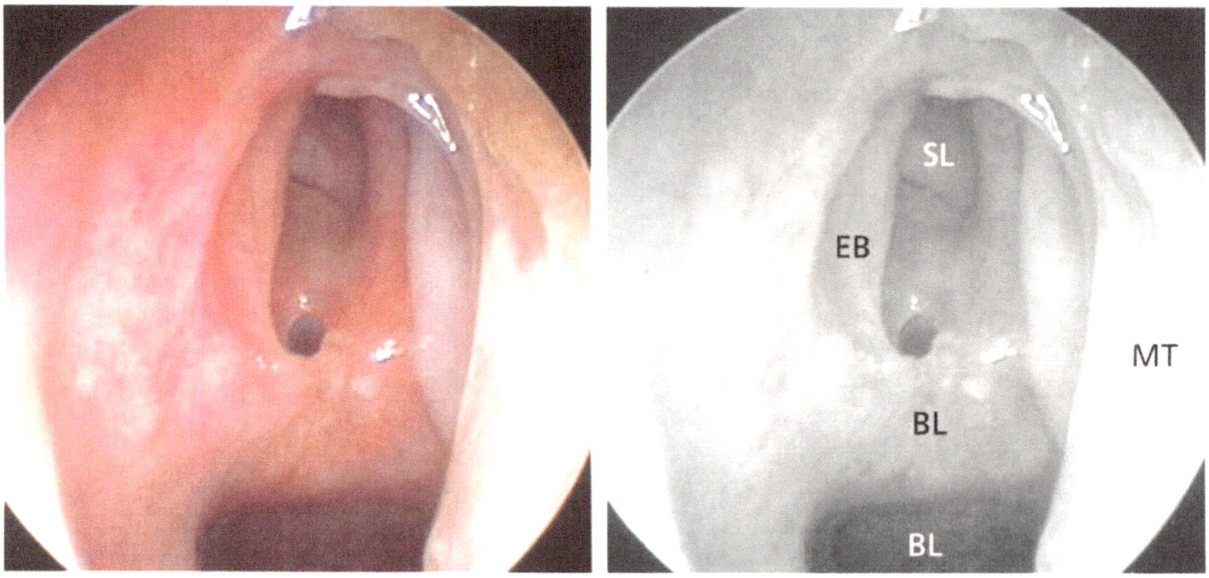

Fig. 7.11. Right side, zero degree lens, post-operative, horizontal and vertical portions of basal lamella, secretion from frontal sinus.

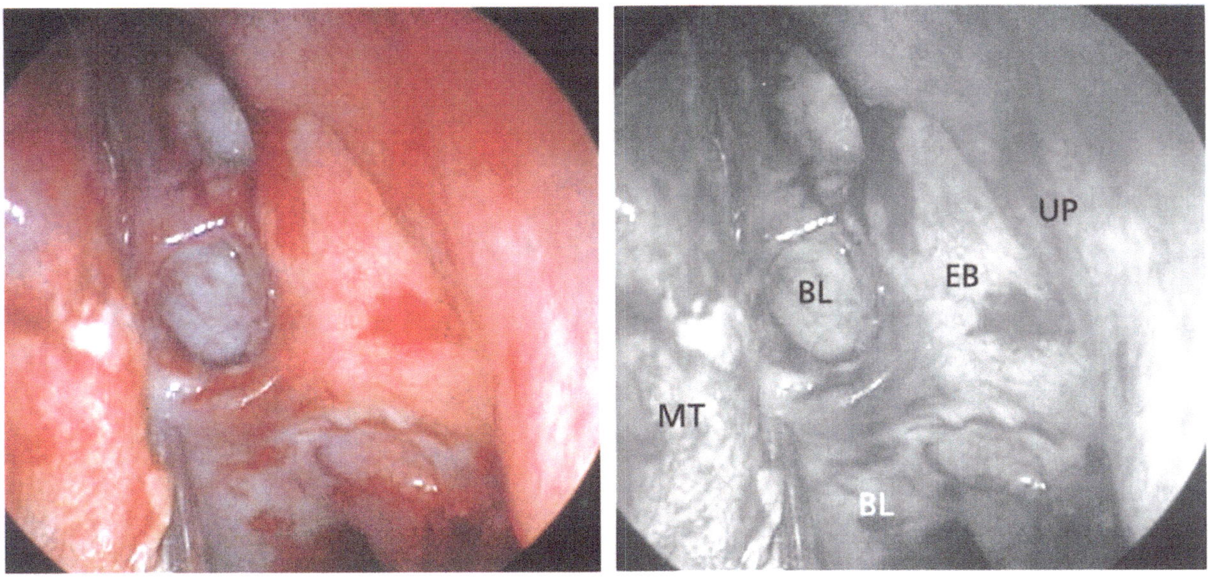

Fig. 7.12. Left side, zero degree lens, horizontal and vertical portions of basal lamella.

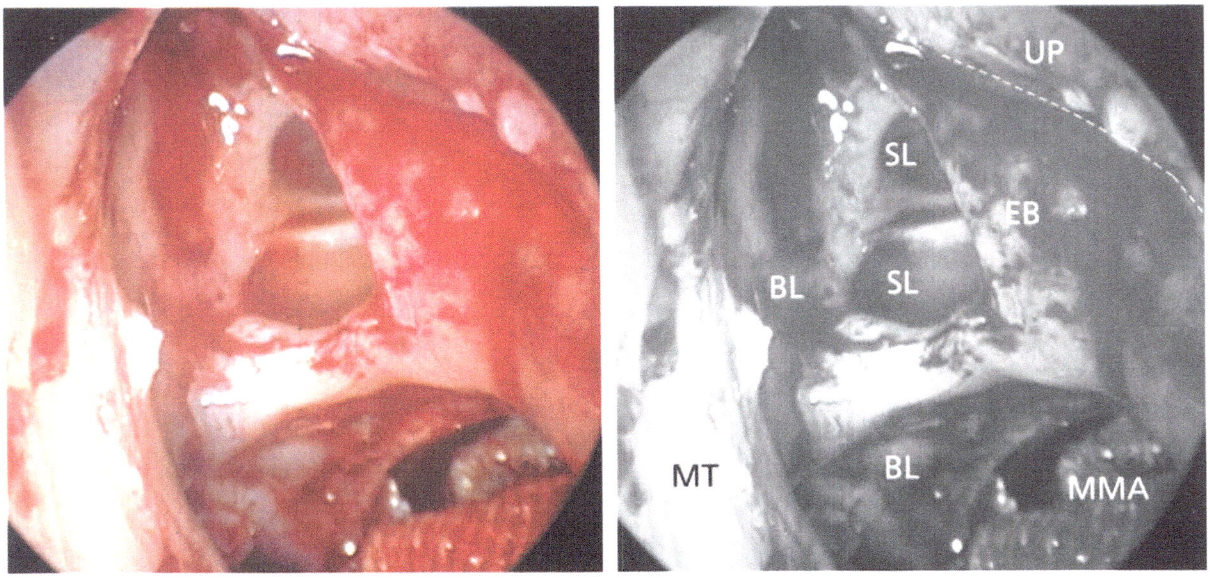

Fig. 7.13. Left side, zero degree lens, sponge in middle meatal antrostomy, horizontal and vertical portions of basal lamella with lateral pneumatization.

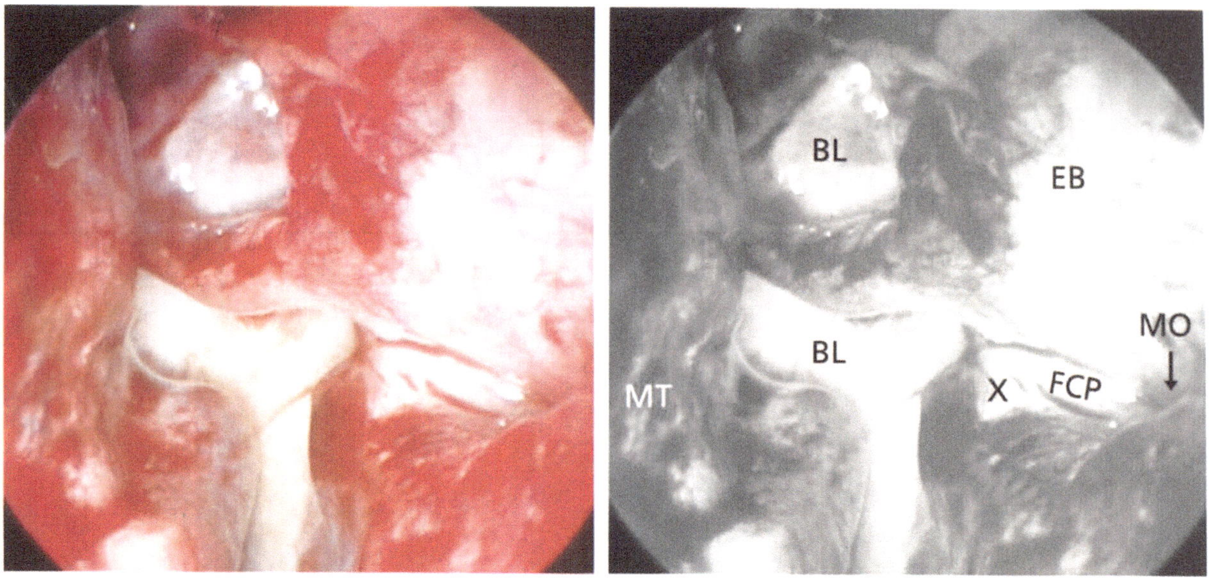

Fig. 7.14. Left side, zero degree lens, horizontal and vertical portions of basal lamella.

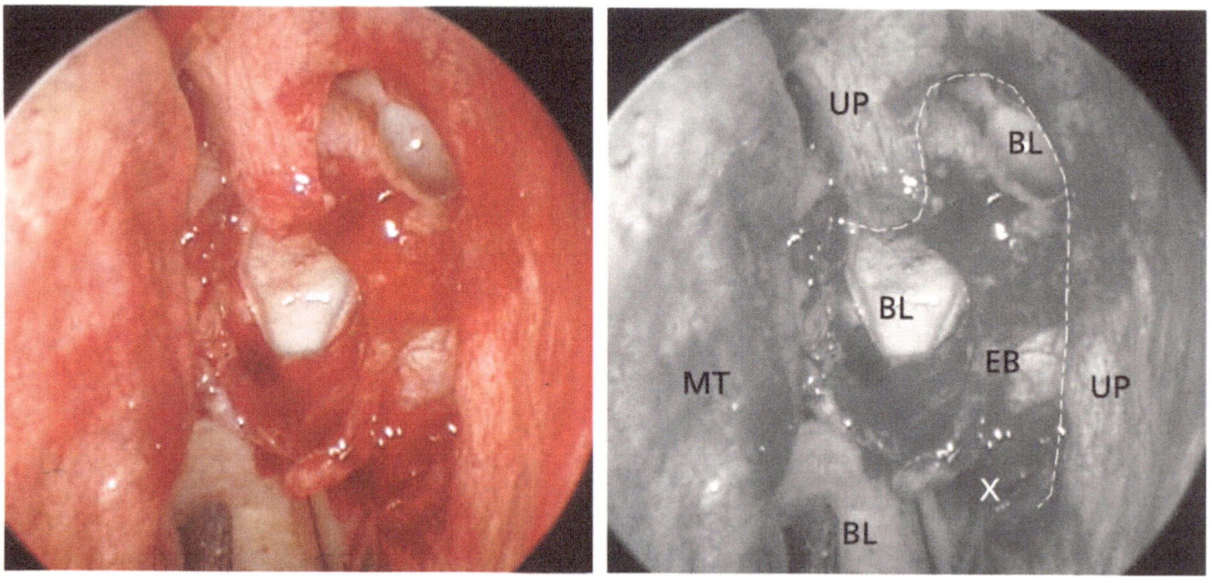

Fig. 7.15. Left side, zero degree lens, horizontal and vertical portions of basal lamella following partial uncinectomy and opening of bulla.

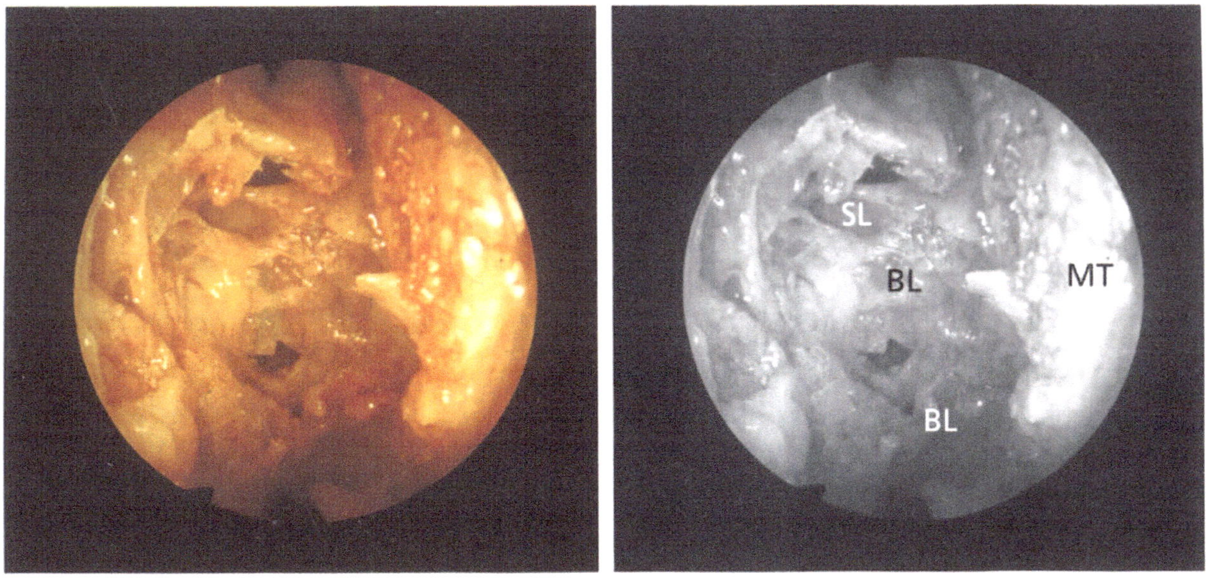

Fig. 7.16. Right side, 30 degree lens, very complex basal lamella with sinus lateralis above.

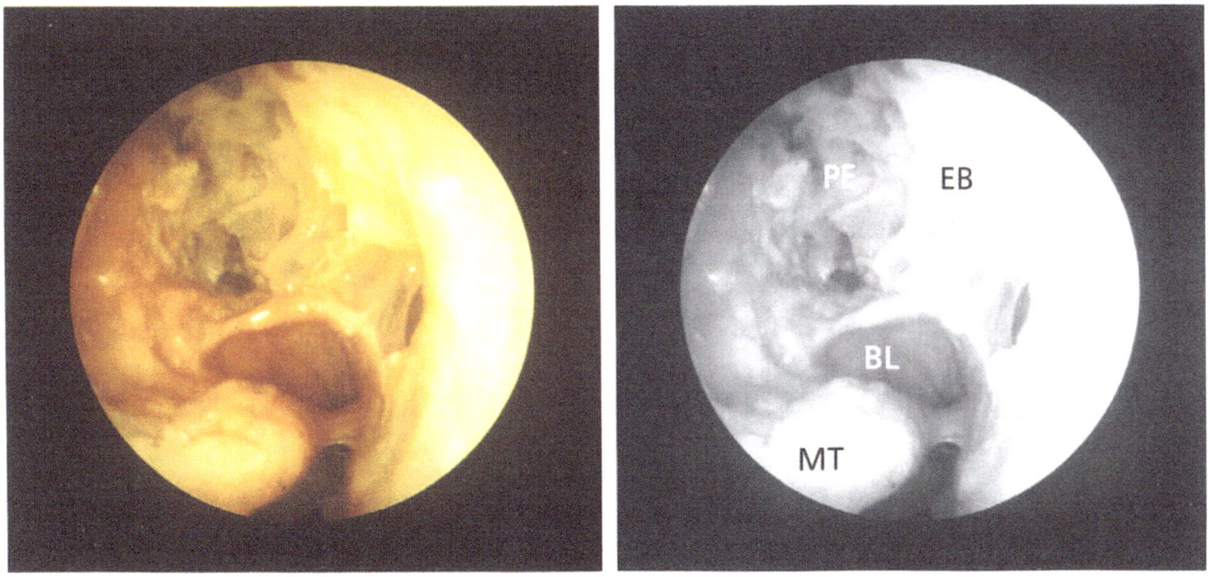

Fig. 7.17. Left side, zero degree lens, horizontal portion intact, vertical portion has been opened.

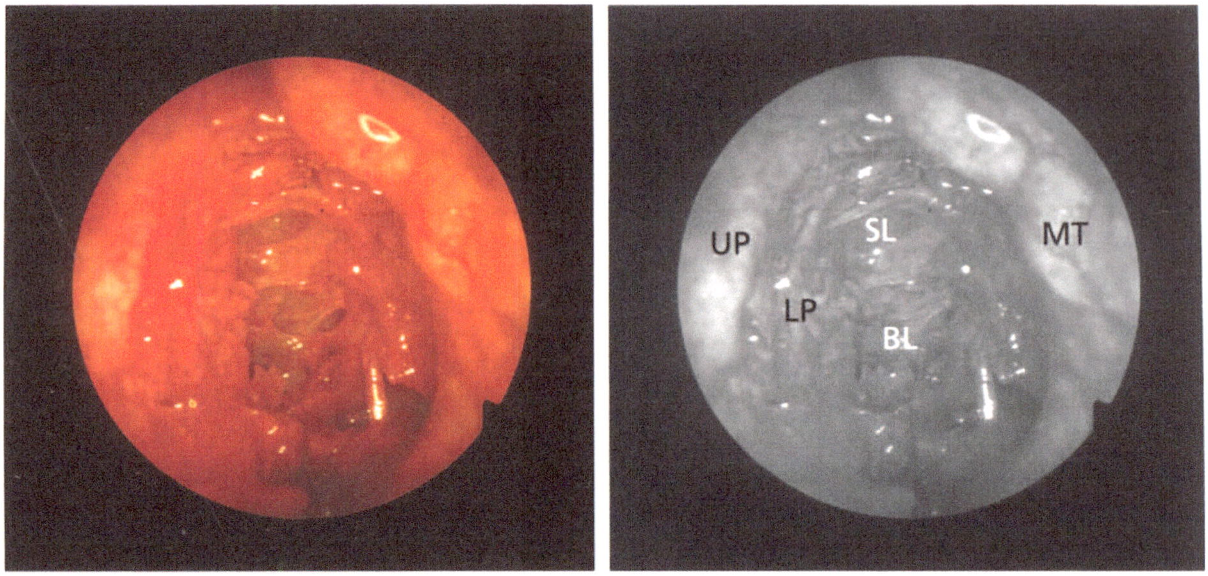

Fig. 7.18. Right side, zero degree lens, complicated basal lamella in its vertical portion with sinus lateralis above.

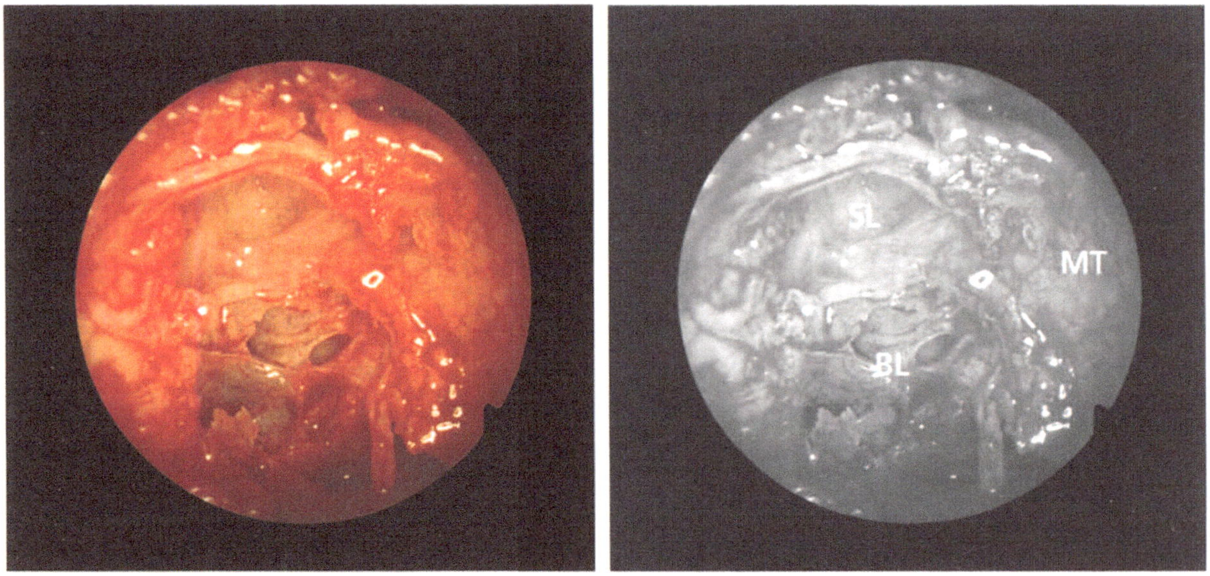

Fig. 7.19. Right side, zero degree lens, close-up of Fig. 7.18.

8. The maxillary sinus ostium, final common pathway, and exit of infundibulum

The maxillary sinus ostium is found in the floor of the infundibulum and is in an oblique plane with respect to the midline. It sometimes is in a near horizontal plane. The drainage from the sinus into the infundibulum tracks posteriorly along a groove (final common pathway) before exiting the infundibulum at the most inferior portion of the hiatus semilunaris between the tail of the uncinate process and the inferior portion of the ethmoidal bulla. In addition to the variation in the plane, the ostium also varies as to its location within the floor of the infundibulum, its size, and the direction that it takes from the maxillary sinus. For the endoscopist, the most reliable anatomical landmark is the exit of the infundibulum. A visual track anteriorly along the final common pathway will lead to the oblique maxillary sinus ostium dropping off inferiorly and laterally into the maxillary sinus.

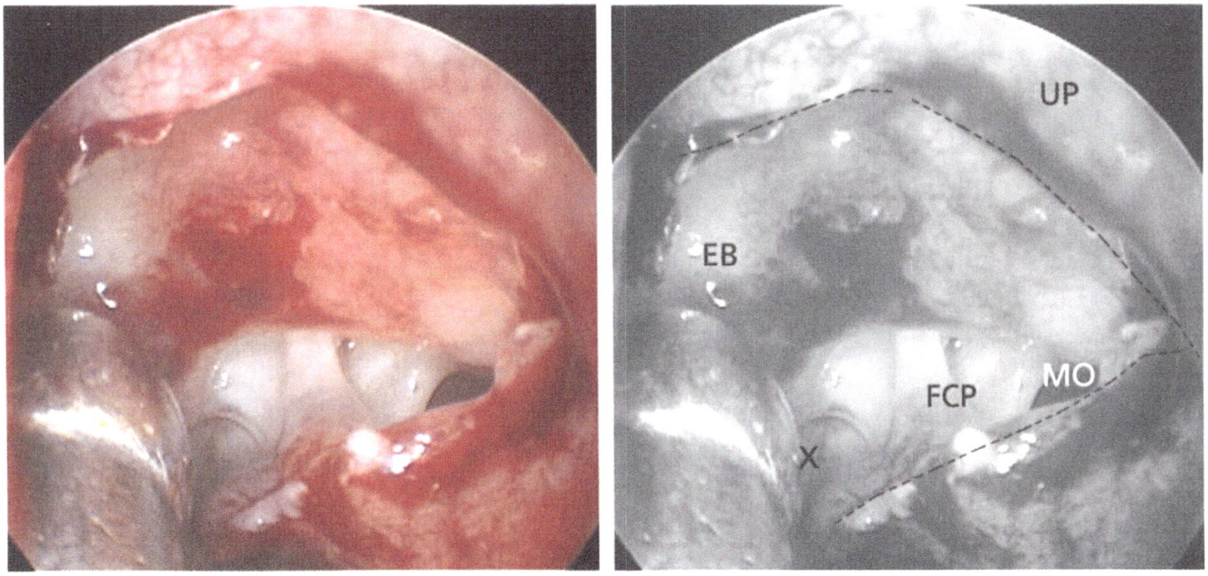

Fig. 8.1. Left side, 30 degree lens, maxillary sinus ostium and final common pathway as seen through uncinate window.

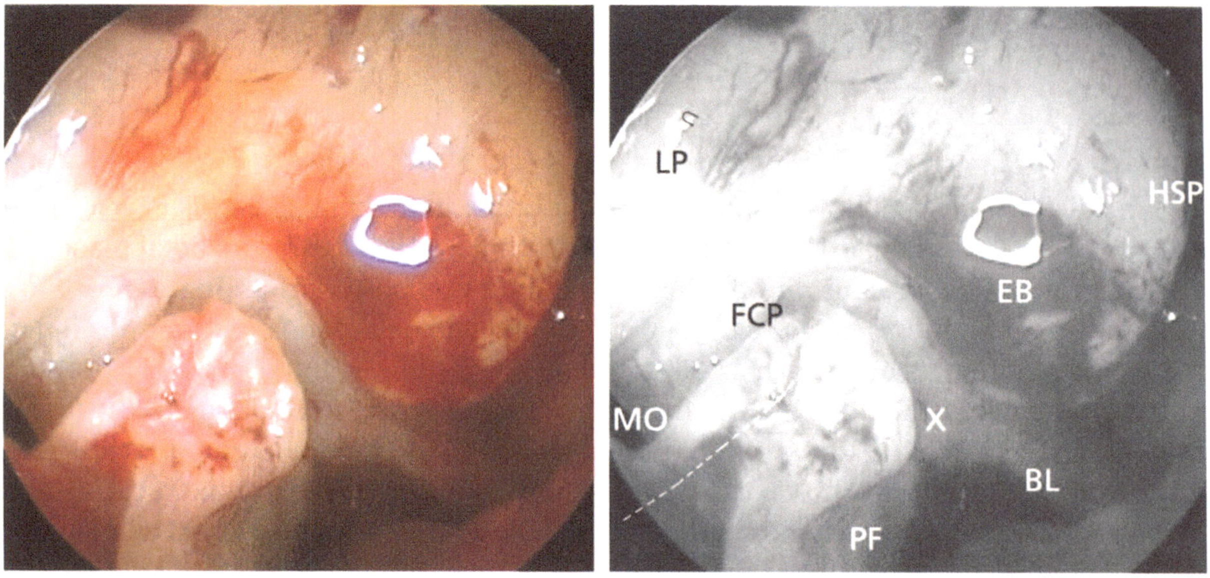

Fig. 8.2. Right side, 30 degree lens, maxillary sinus ostium and final common pathway and exit of infundibulum with overhanging ethmoidal bulla; tail of uncinate process has not been removed.

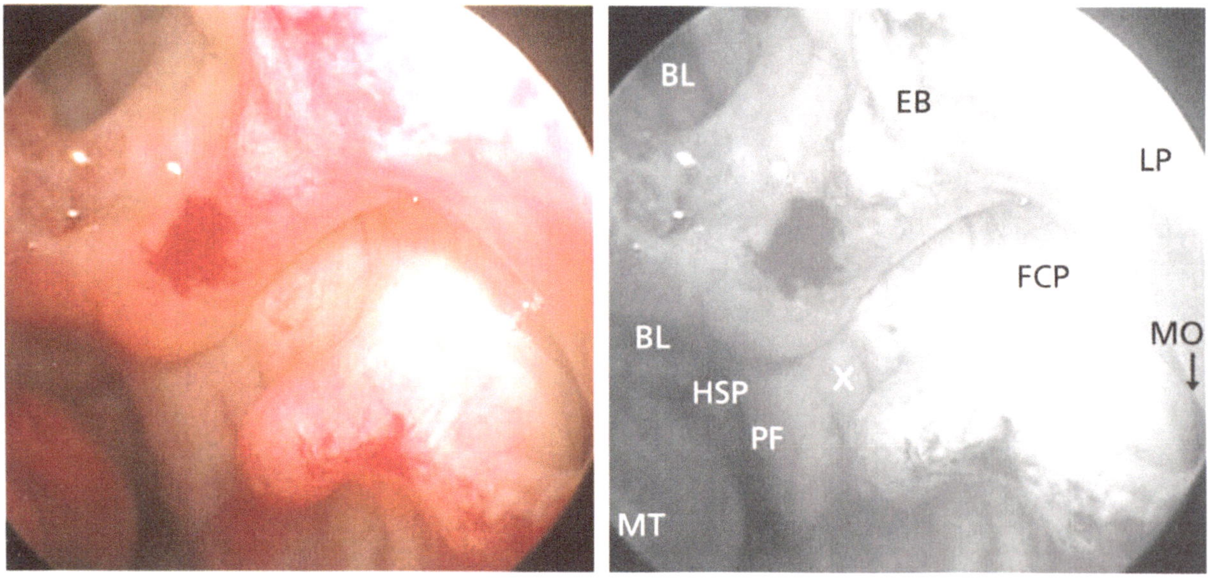

Fig. 8.3. Left side, 30 degree lens, maxillary sinus ostium and arching final common pathway with exit onto posterior fontanelle; open ethmoidal bulla, vertical and horizontal portions of basal lamella, and lamina papyracea are visible. Tail of uncinate has not been removed.

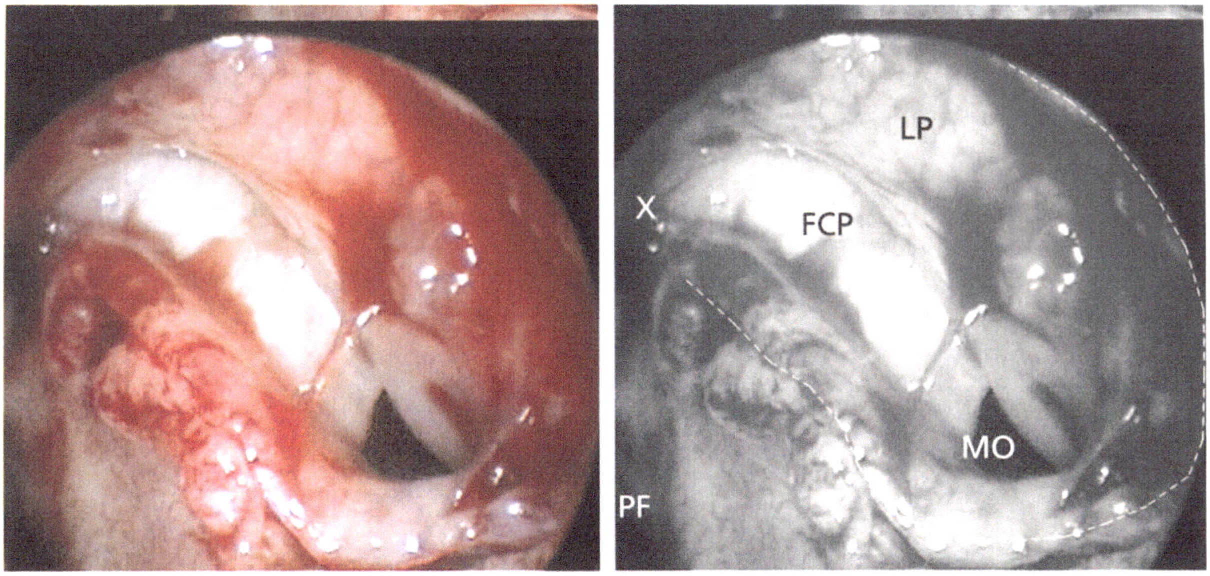

Fig. 8.4. Left side, 30 degree lens, maxillary sinus ostium with Haller cell; long final common pathway and exit onto posterior fontanelle; tail of uncinate has been removed with apposition of nasal mucosa and infundibular mucous membrane.

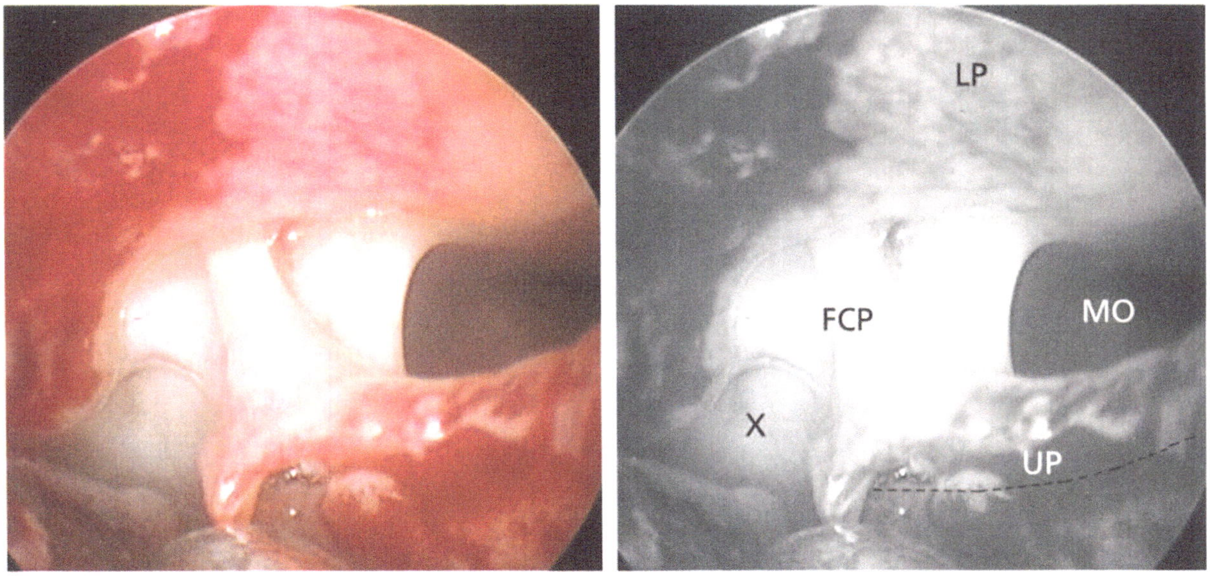

Fig. 8.5. Left side, 30 degree lens, maxillary sinus ostium showing a floor of orbit, very short final common pathway, and exit; tail of uncinate has been removed.

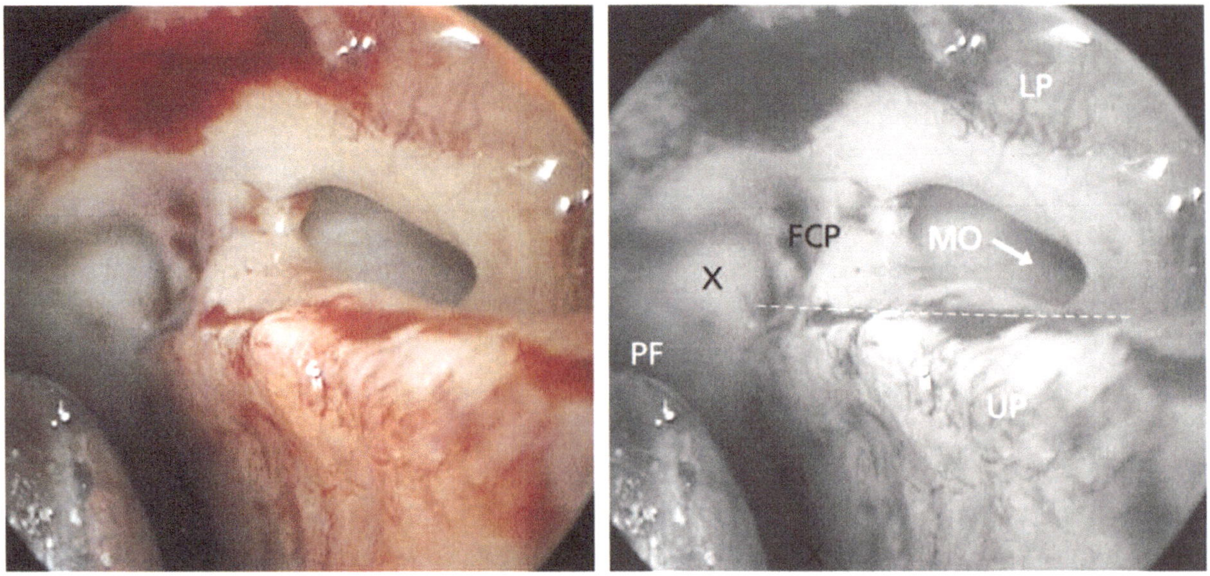

Fig. 8.6. Left side, 30 degree lens, vestibule to maxillary sinus ostium, mid-infundibular floor location; tail of uncinate has been removed.

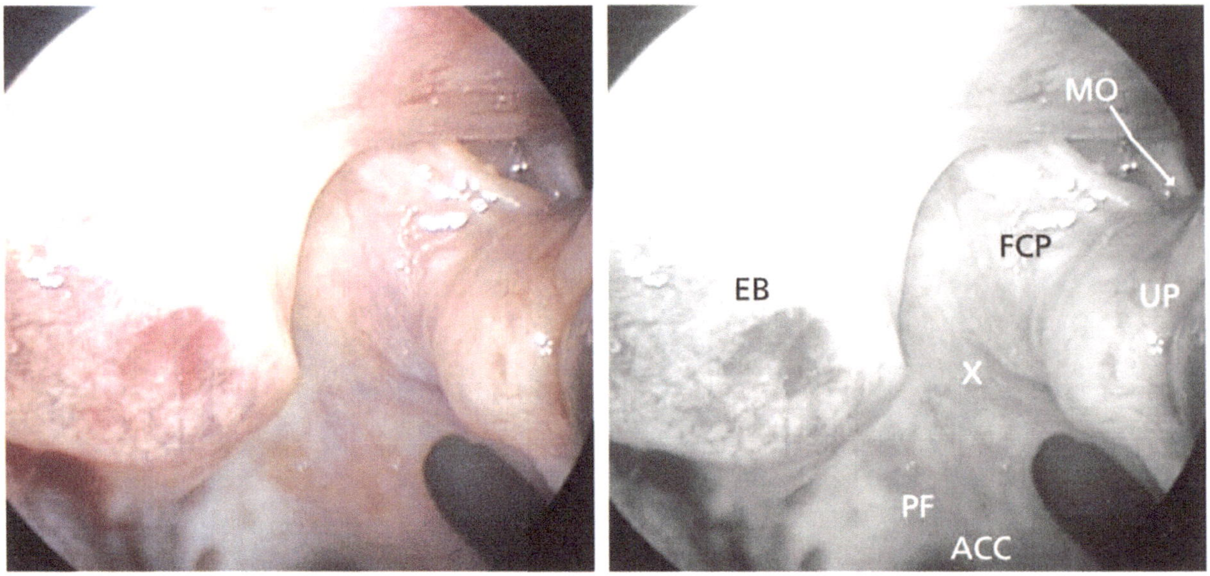

Fig. 8.7. Left side, zero degree lens, maxillary sinus ostium in posterior location with short final common pathway and exit beneath an overhanging ethmoidal bulla; accessory ostium and posterior fontanelle.

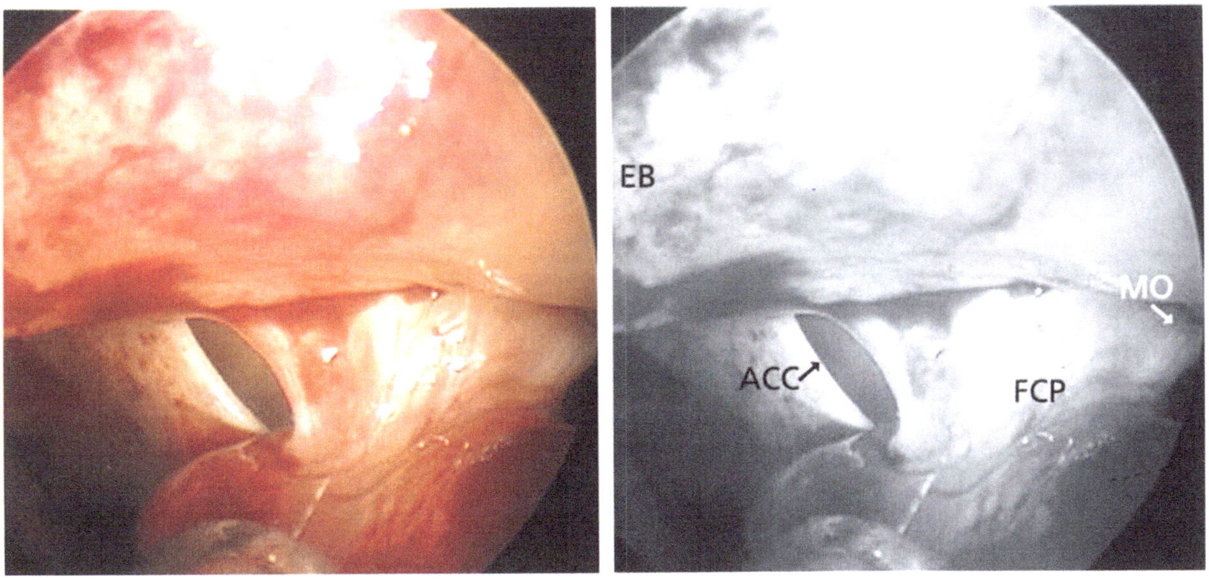

Fig. 8.8. Left side, 30 degree lens, maxillary sinus ostium in oblique plane, accessory ostium within final common pathway in parasagittal plane.

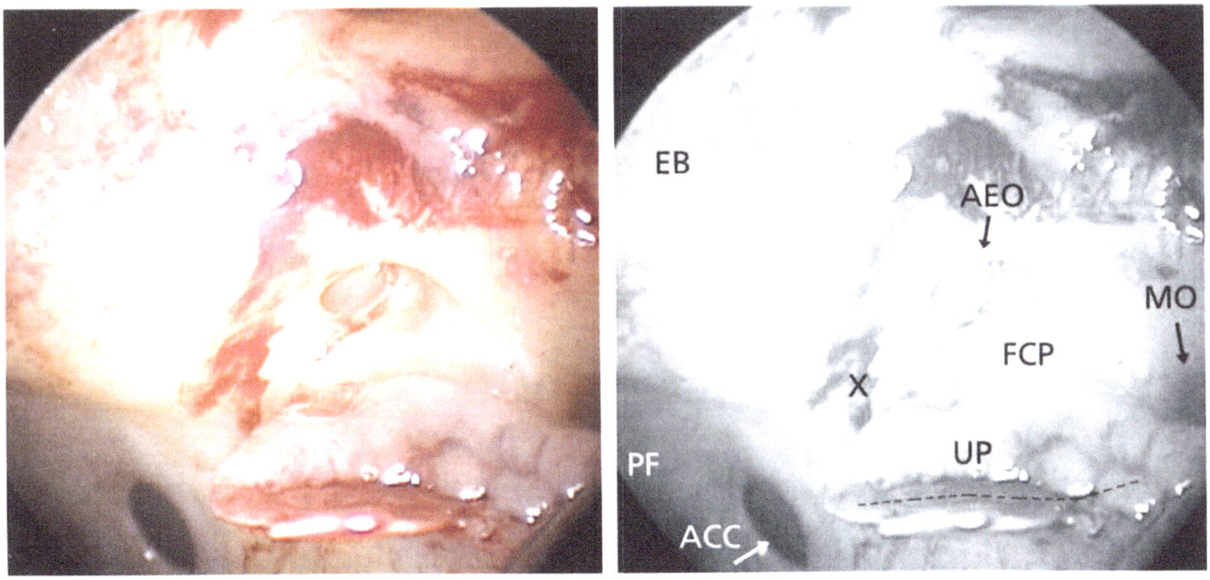

Fig. 8.9. Left side, 30 degree lens, maxillary sinus ostium, ostium for anterior ethmoid cell, and accessory ostium of posterior fontanelle.

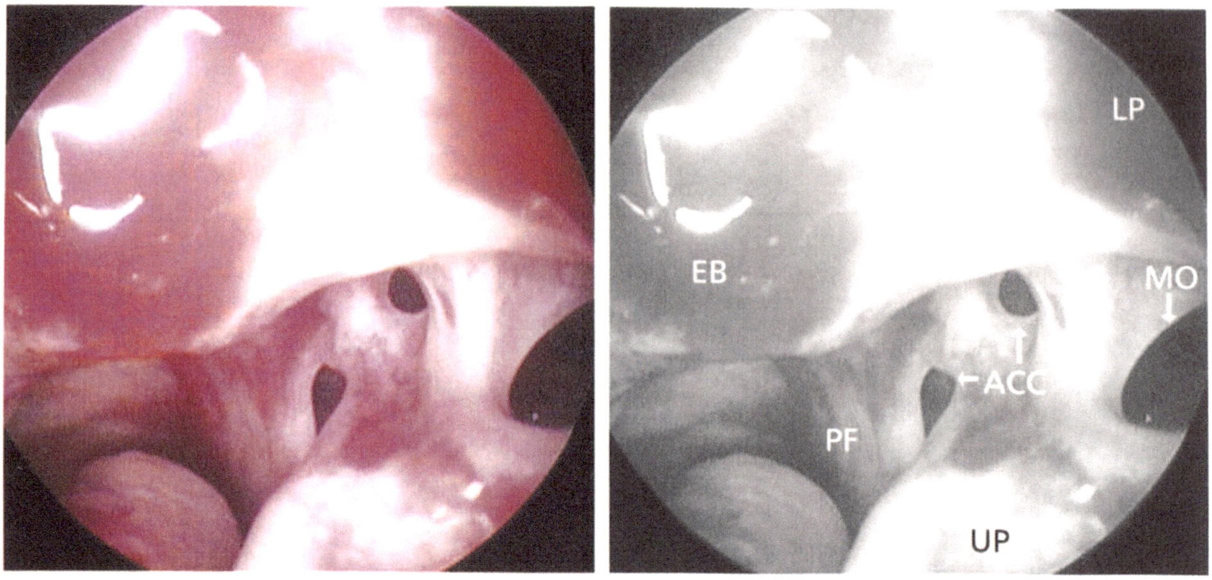

Fig. 8.10. Left side, 30 degree lens, maxillary sinus ostium and two accessory ostia within the final common pathway.

9. The sinus lateralis

The sinus lateralis is a space or recess of varying proportion behind and/or superior to the ethmoidal bulla. It may be recognized as an apparent pneumatization or indentation of any part of the basal lamella and may extend to the base of the skull. The sinus lateralis opens medially toward the middle turbinate.

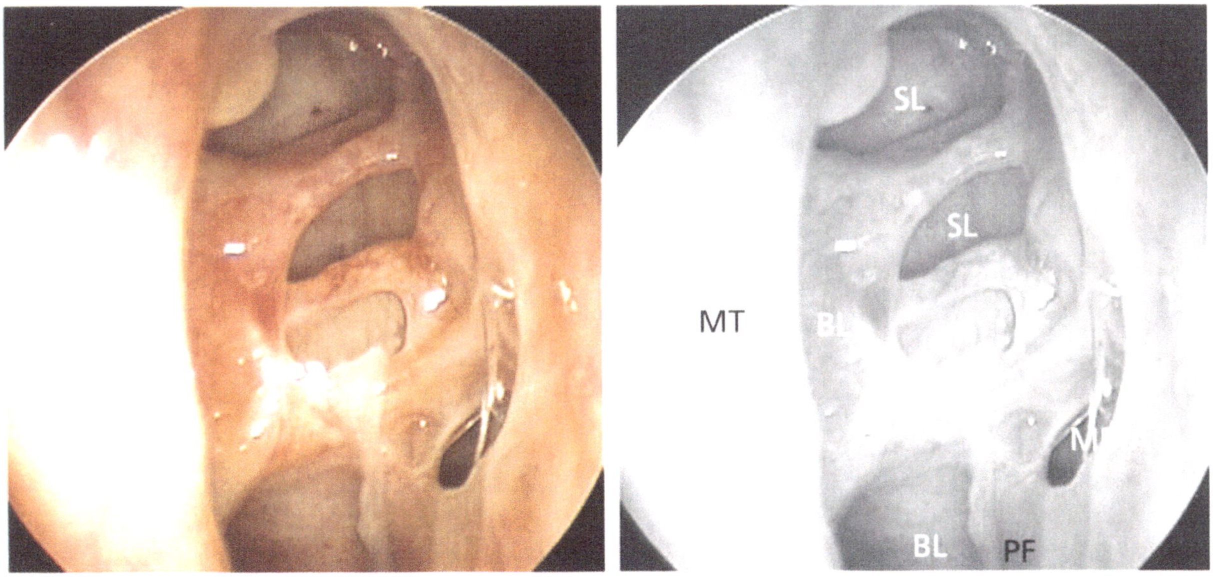

Fig. 9.1. Left side, zero degree lens, pneumatization of basal lamella in a post-operative patient; i.e., sinus lateralis.

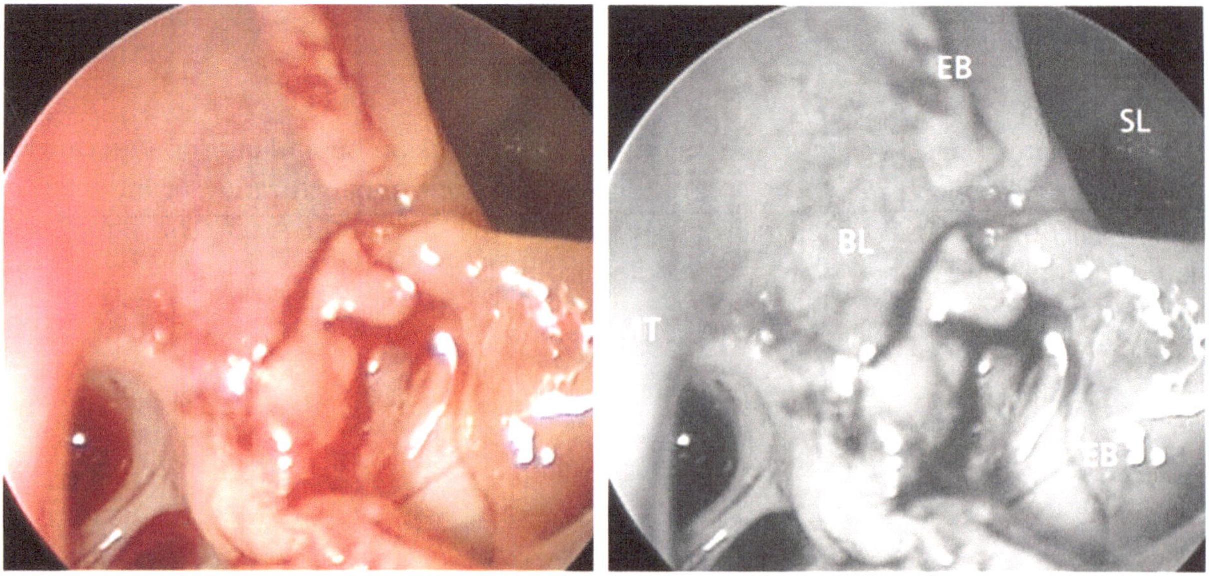

Fig. 9.2. Left side, zero degree lens, basal lamella and sinus lateralis behind the ethmoidal bulla.

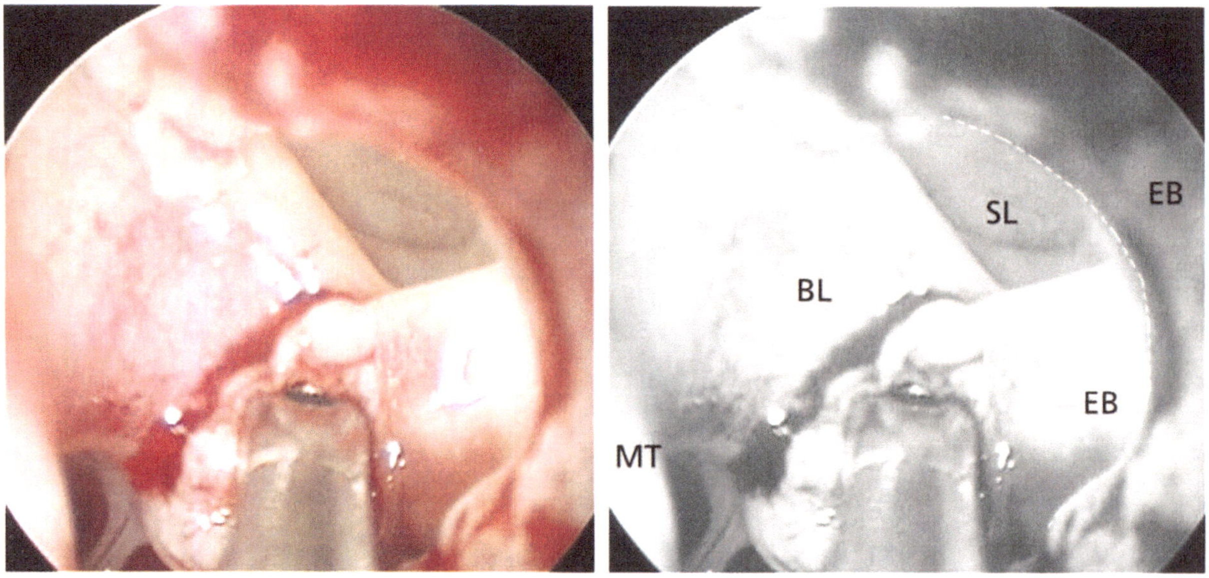

Fig. 9.3. Left side, zero degree lens, entering basal lamella, sinus lateralis above.

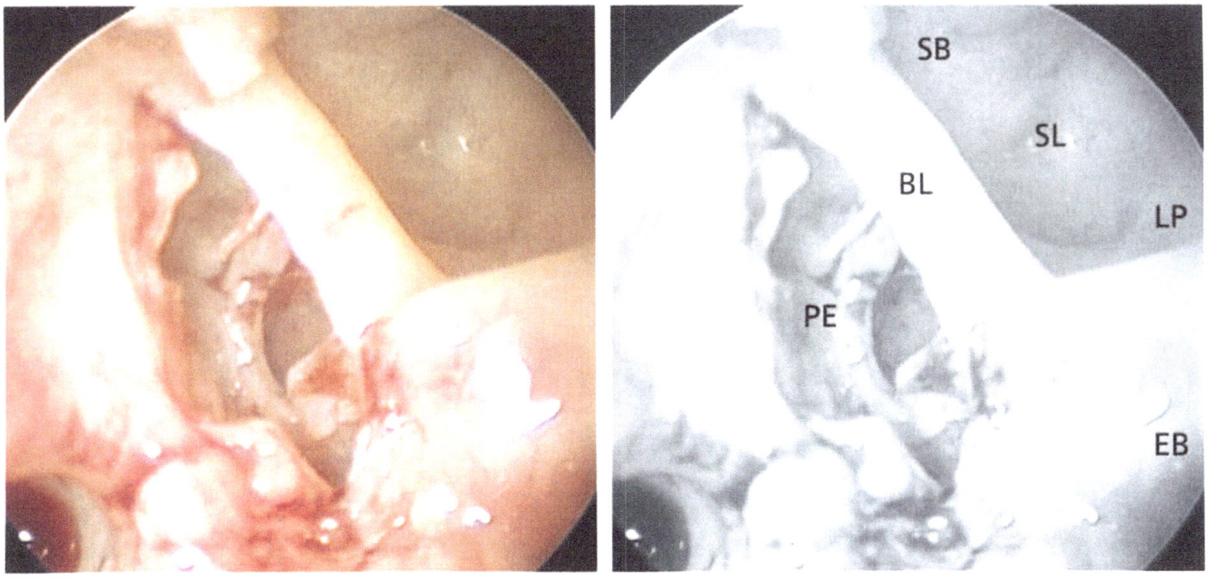

Fig. 9.4. Left side, zero degree lens, posterior ethmoid below, sinus lateralis above.

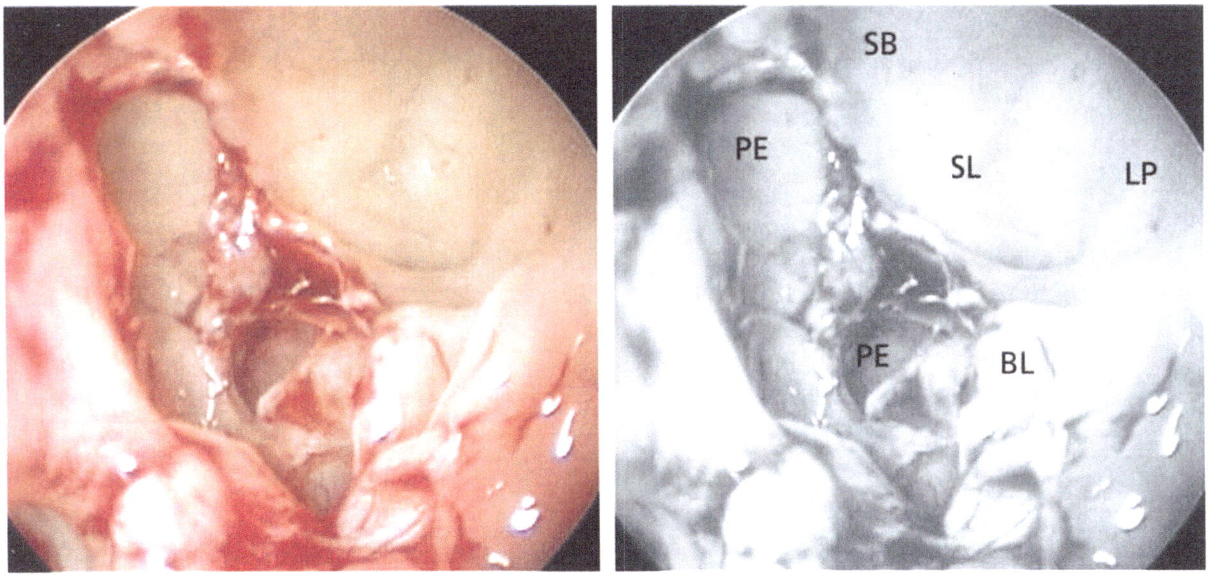

Fig. 9.5. Left side, zero degree lens, same patient, posterior ethmoid and sinus lateralis are one.

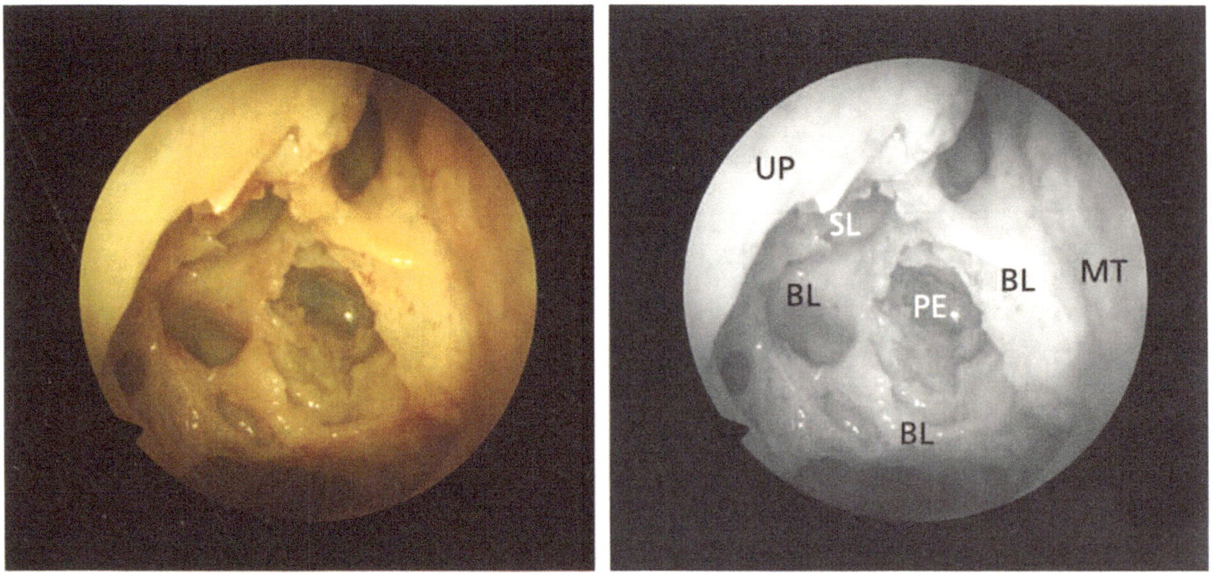

Fig. 9.6. Right side, zero degree lens, posterior ethmoid and sinus lateralis above; ethmoidal bulla along side lamina papyracea below.

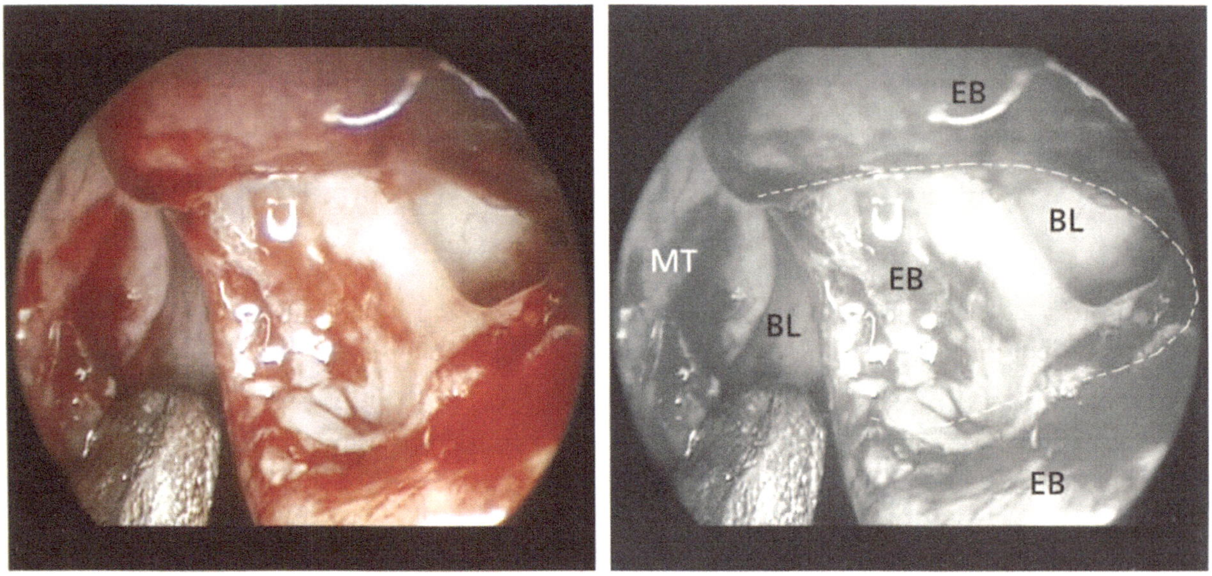

Fig. 9.7. Left side, zero degree lens, posterior wall of ethmoidal bulla; partially pneumatized; recess between it and the basal lamella is sinus lateralis.

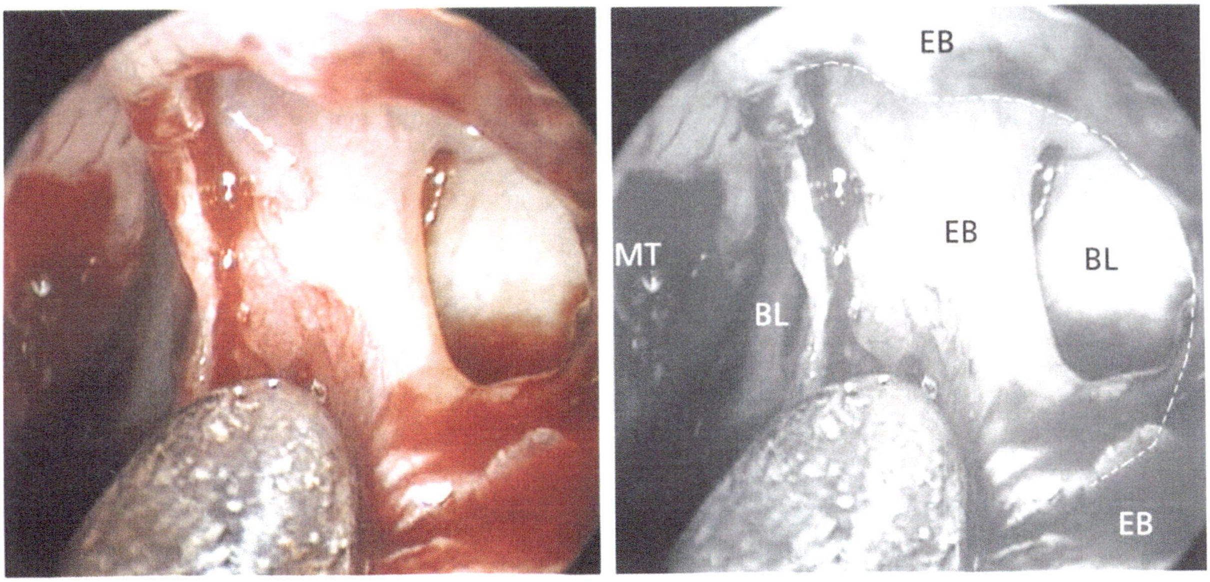

Fig. 9.8. Left side, zero degree lens, same as Fig. 9.7.

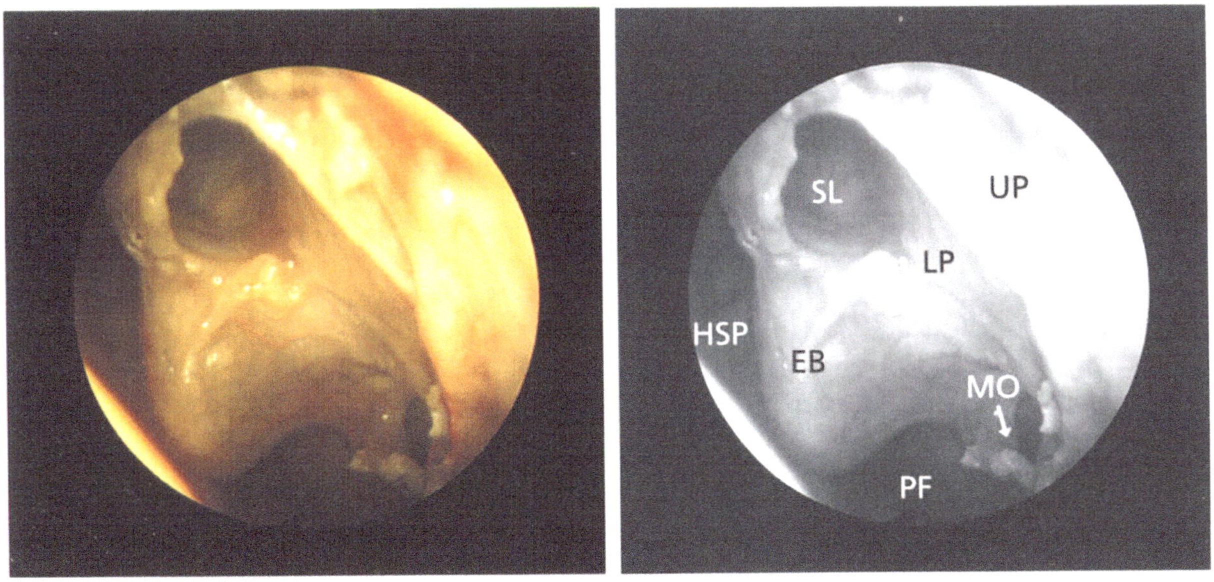

Fig. 9.9. Left side, 30 degree lens, sinus lateralis above ethmoidal bulla.

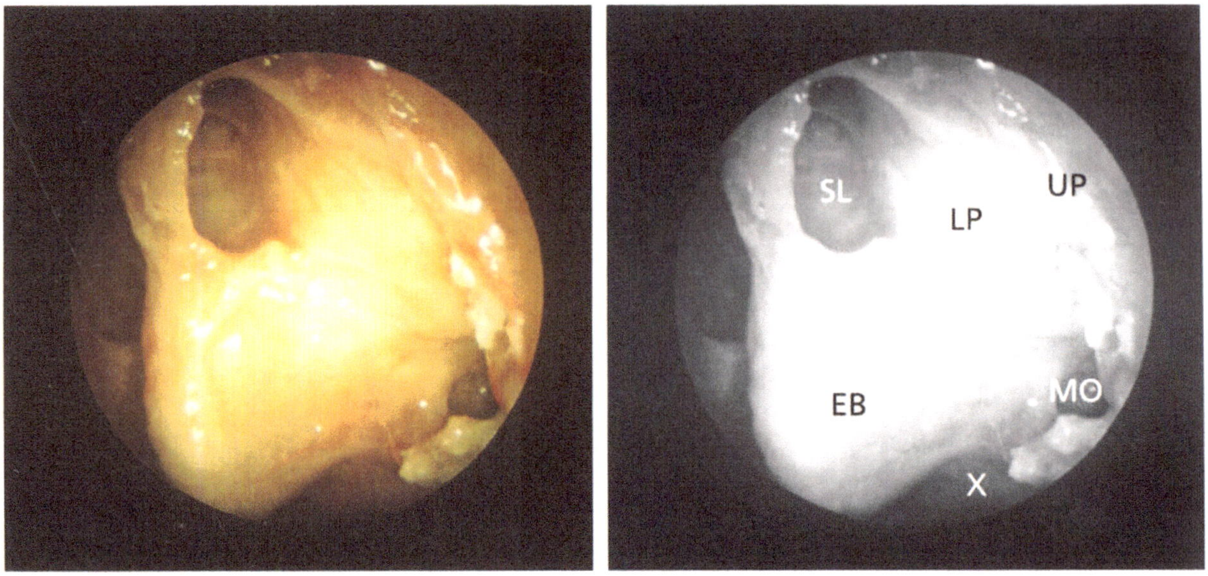

Fig. 9.10. Left side, 70 degree lens, same as Fig. 9.9.

10. The agger nasi cell

The agger nasi is an external feature of the ethmoid complex. It is visible to the endoscopist immediately above the anterior attachment of the middle turbinate. The agger nasi cell itself is the most anterior superior cell of the ethmoid complex. The cell and its posterior medial wall constitute the fourth critical anatomic landmark for the endoscopist.

The drainage for the frontal sinus is, in most operative instances, immediately behind the posterior-medial wall of the agger nasi cell. The dome of the cell is more or less the floor of the frontal sinus. Variations include a dehiscence of the posterior wall, an extension of the cell into the frontal sinus, and a direct entry of the frontal sinus into the cell. Visualization posterior and medial to the agger nasi cell may reveal multiple openings depending upon the pattern of frontal pneumatization. Being the most anterior superior cell of the ethmoid complex, drainage is inferior. The cell is continuous with the lamina papyracea laterally.

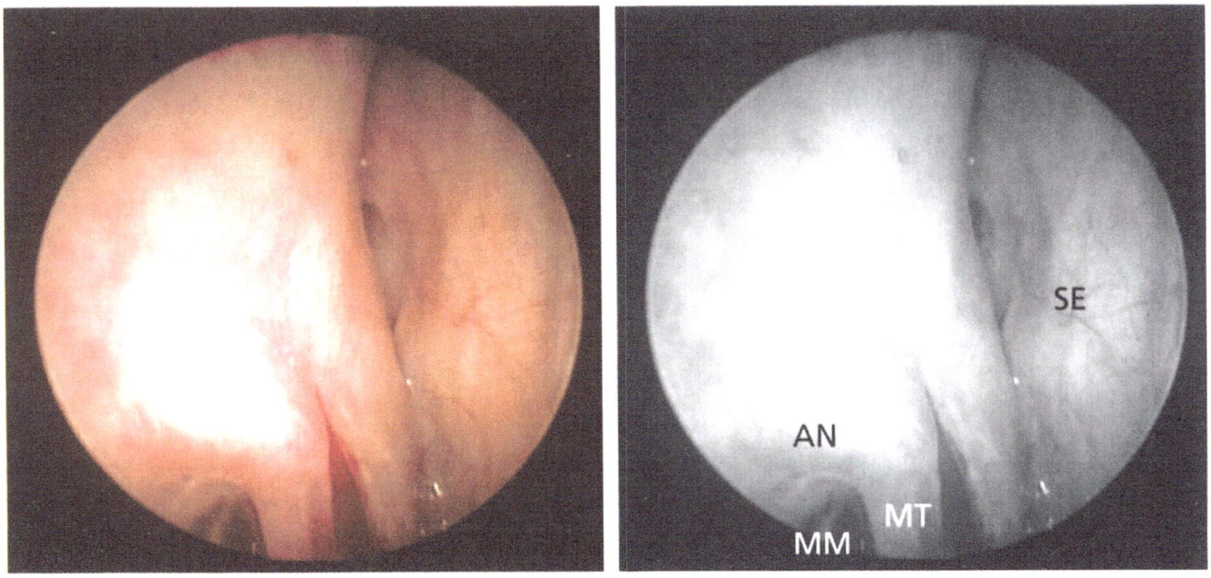

Fig. 10.1. Right side, zero degree lens, small agger nasi.

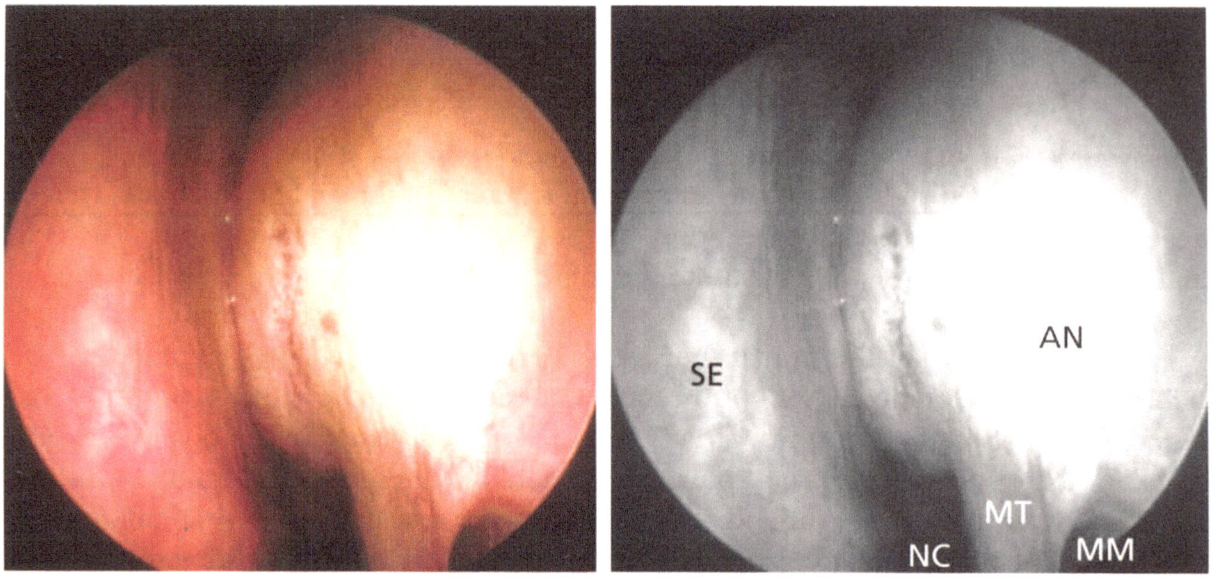

Fig. 10.2. Left side, zero degree lens, very large agger nasi pushing medially into the nasal septum.

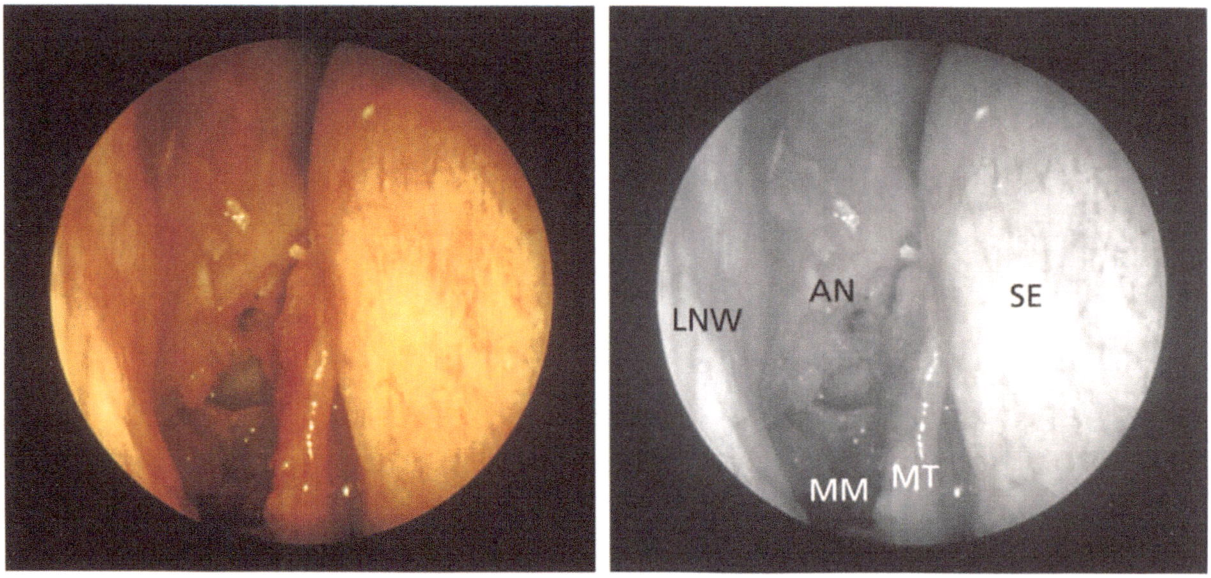

Fig. 10.3. Right side, 30 degree lens, agger nasi.

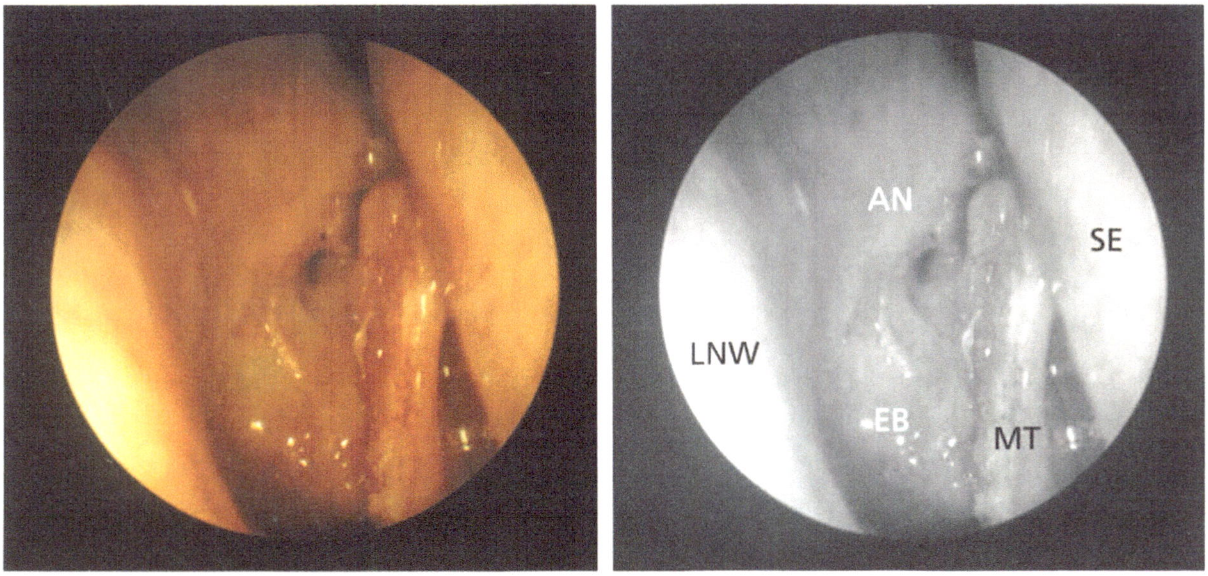

Fig. 10.4. Right side, 30 degree lens, agger nasi.

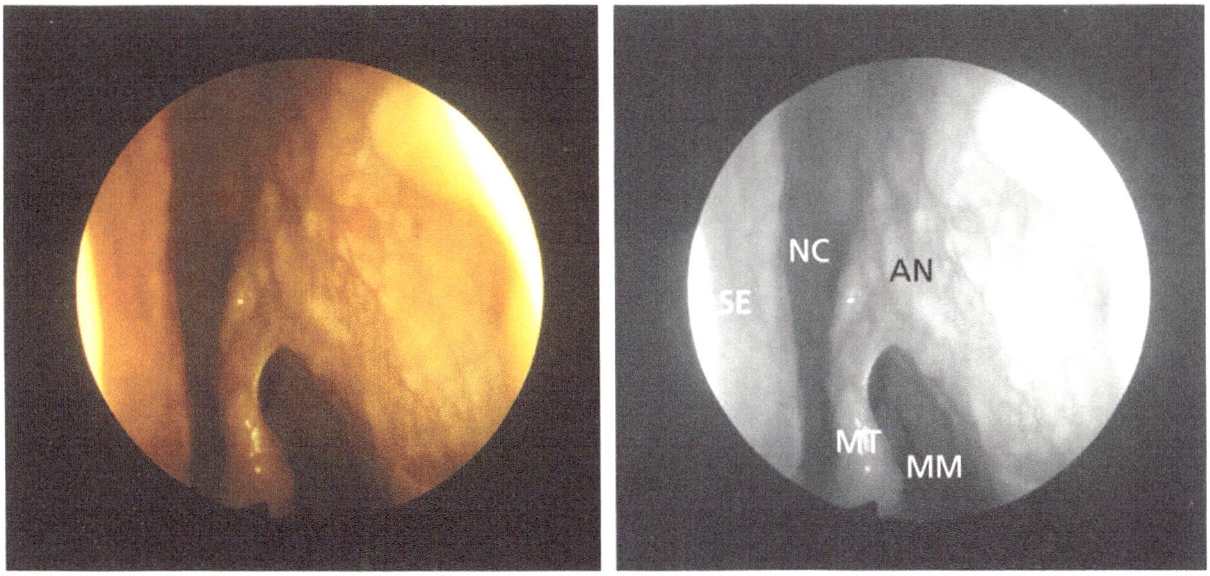

Fig. 10.5. Left side, 30 degree lens, agger nasi.

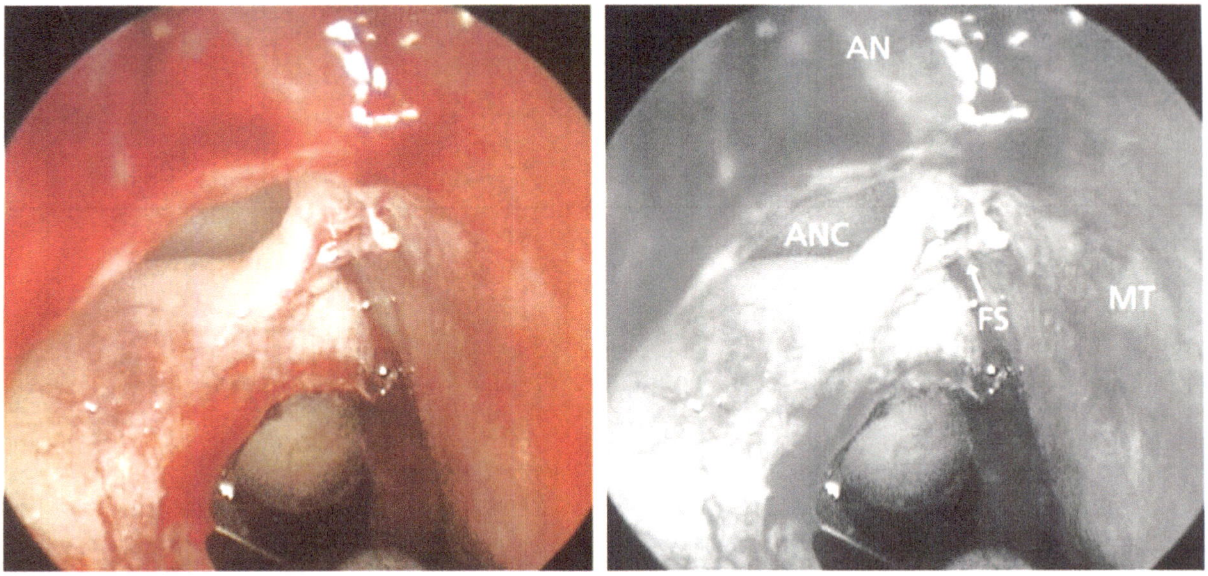

Fig. 10.6. Right side if picture is reversed, zero degree lens, superior residual uncinate from window resection will lead to agger nasi cell.

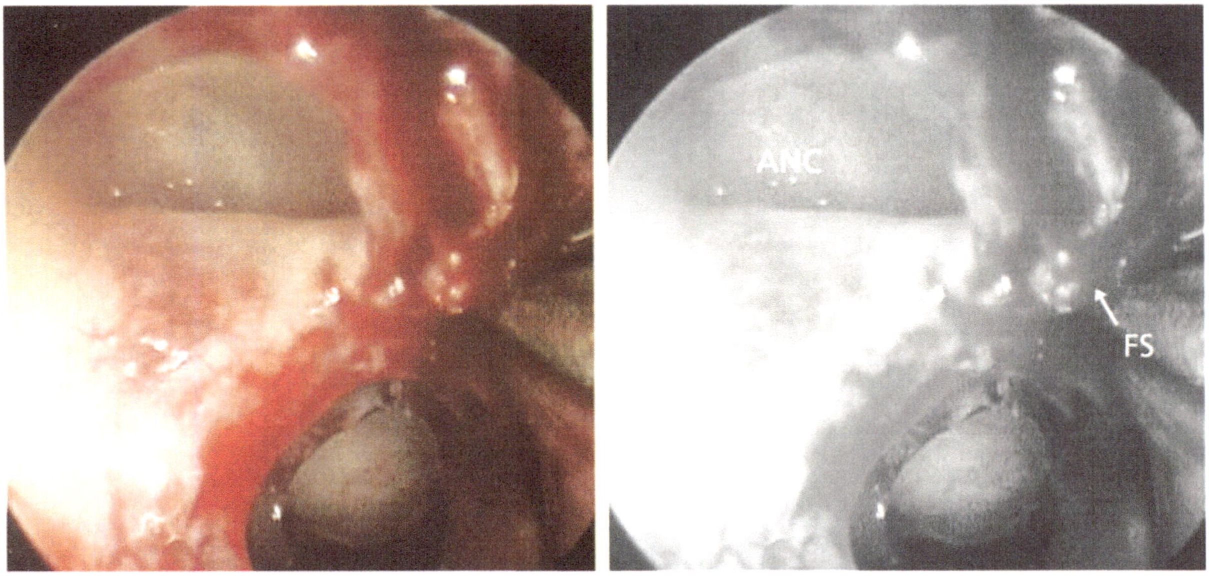

Fig. 10.7. Right side, 30 degree lens, agger nasi cell open.

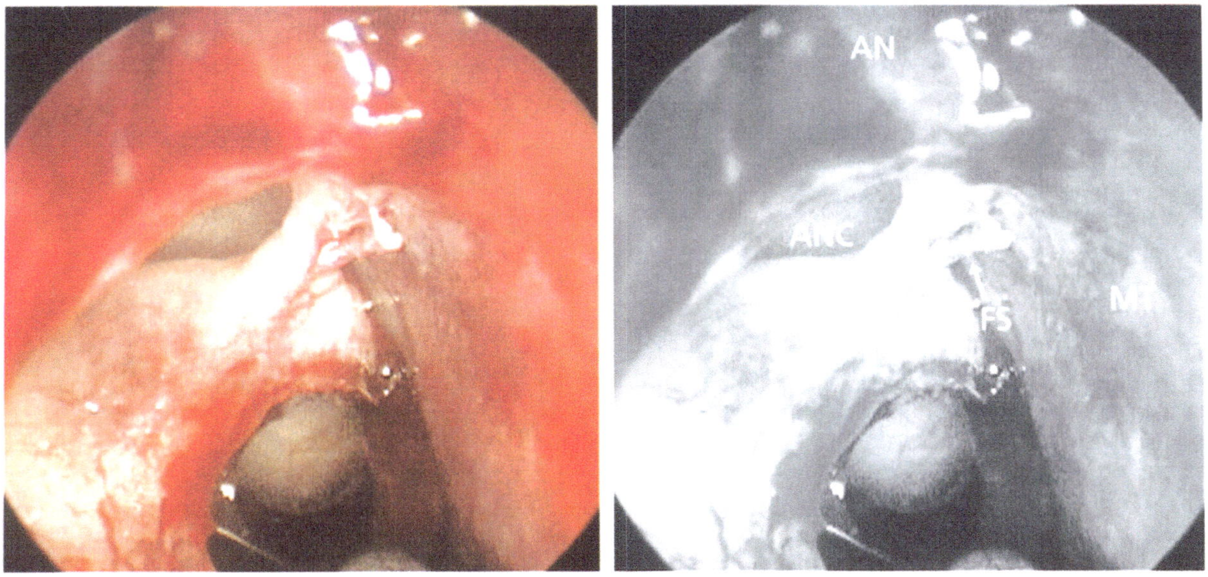

Fig. 10.8. Right side, 30 degree lens, agger nasi cell and view of posterior and medial wall.

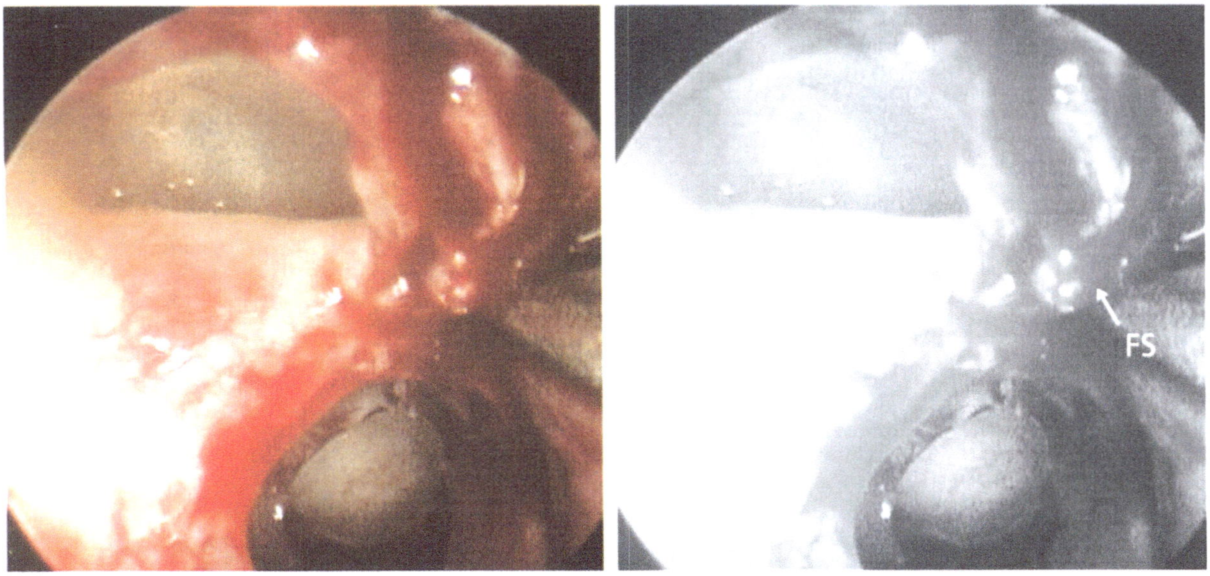

Fig. 10.9. Right side, 30 degree lens, agger nasi cell and dome. Instrument behind posterior and medial wall.

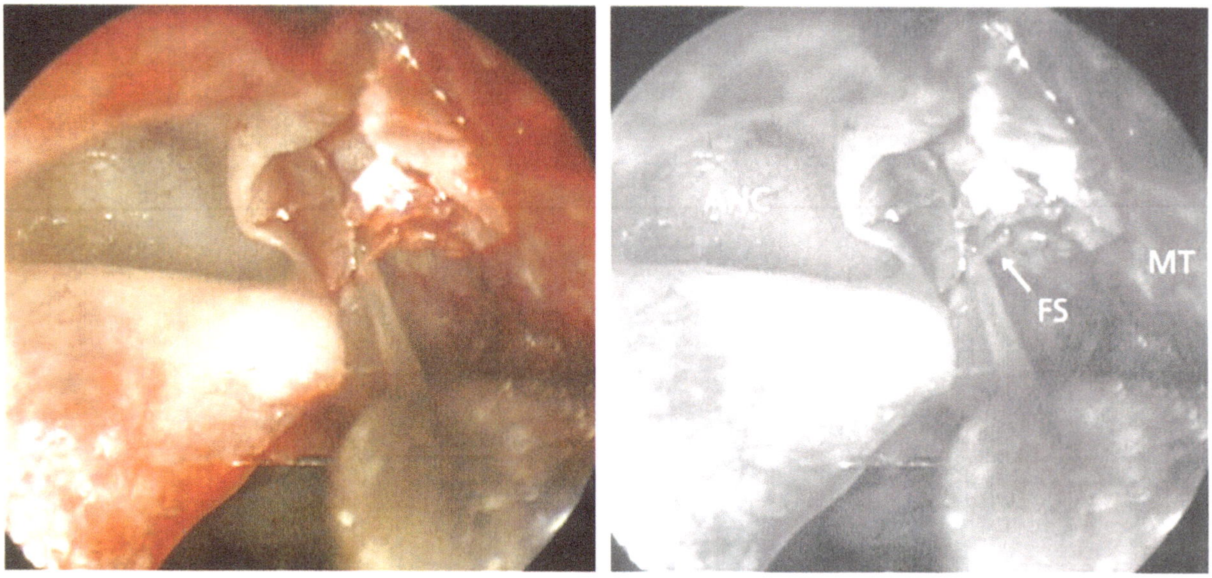

Fig. 10.10. Right side, 30 degree lens, posterior and medial wall partially removed. Secretions from frontal sinus through transition space.

11. The Haller cell

The Haller, or infra-orbital, cell is an extension of anterior ethmoid development beneath the orbit and above the maxillary sinus ostium.

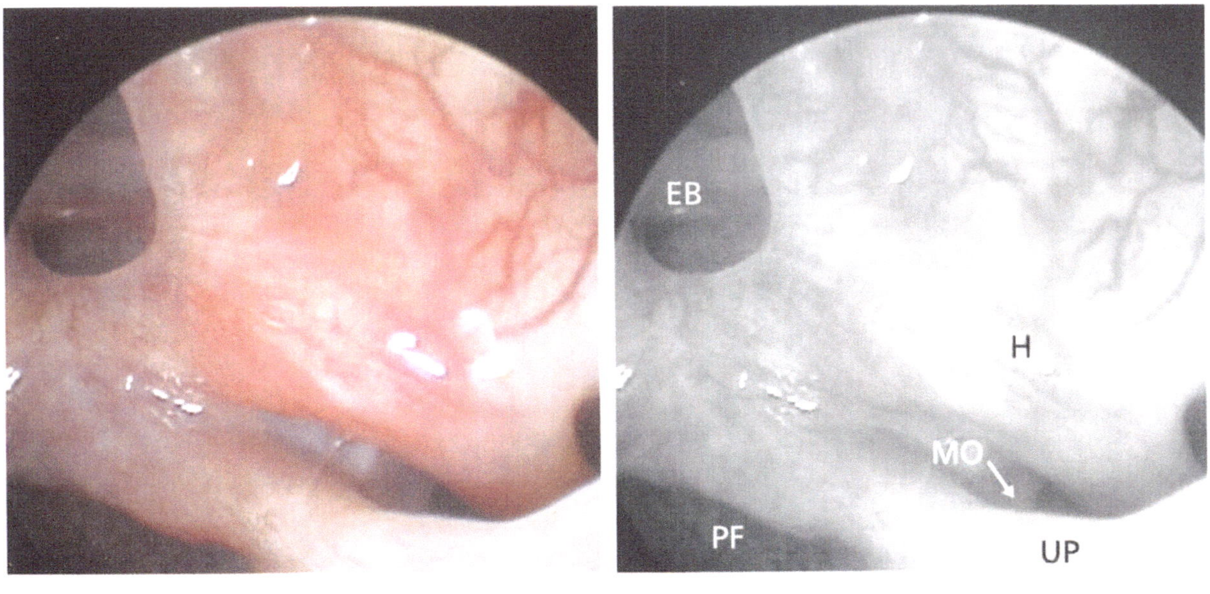

Fig. 11.1. Left side, 30 degree lens, large maxillary sinus ostium with overhanging Haller cell.

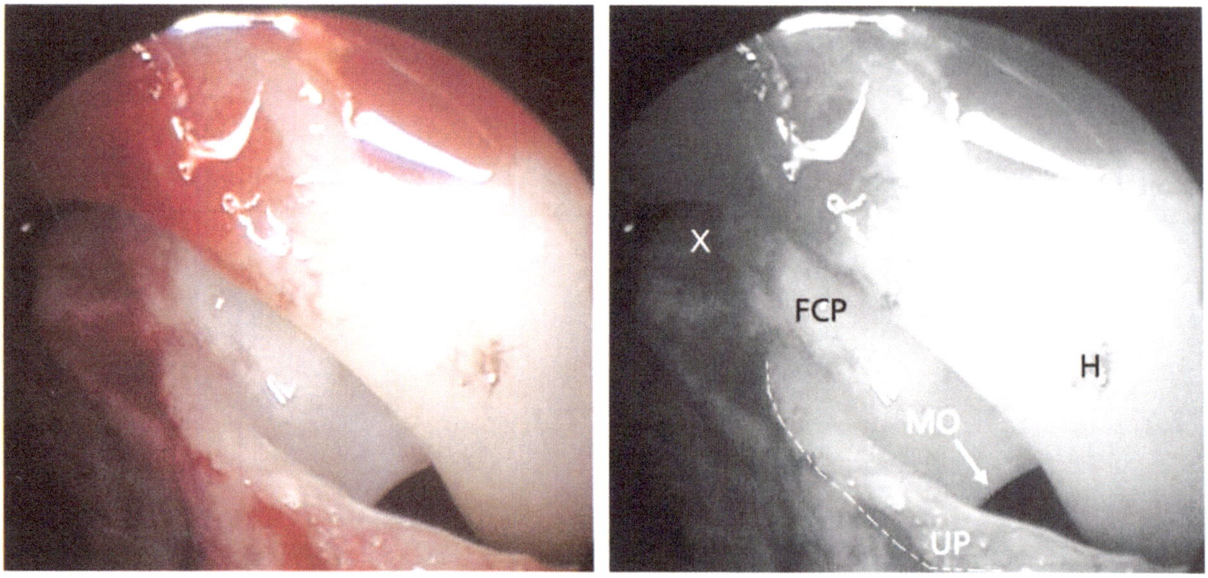

Fig. 11.2. Left side, 30 degree lens, maxillary sinus ostium with overhanging Haller cell.

12. The maxillary sinus

The pneumatized space beneath the orbit is the maxillary sinus. It drains from its cavity into the infundibulum exiting medially at the anterior superior portion of the sinus. The entry into the infundibulum is at its most inferior extent. Accessory ostia are common in the posterior fontanelle, a membranous portion of the medial wall of the sinus. They also occur at the exit, along the final common pathway, and/or in the anterior fontanelle.

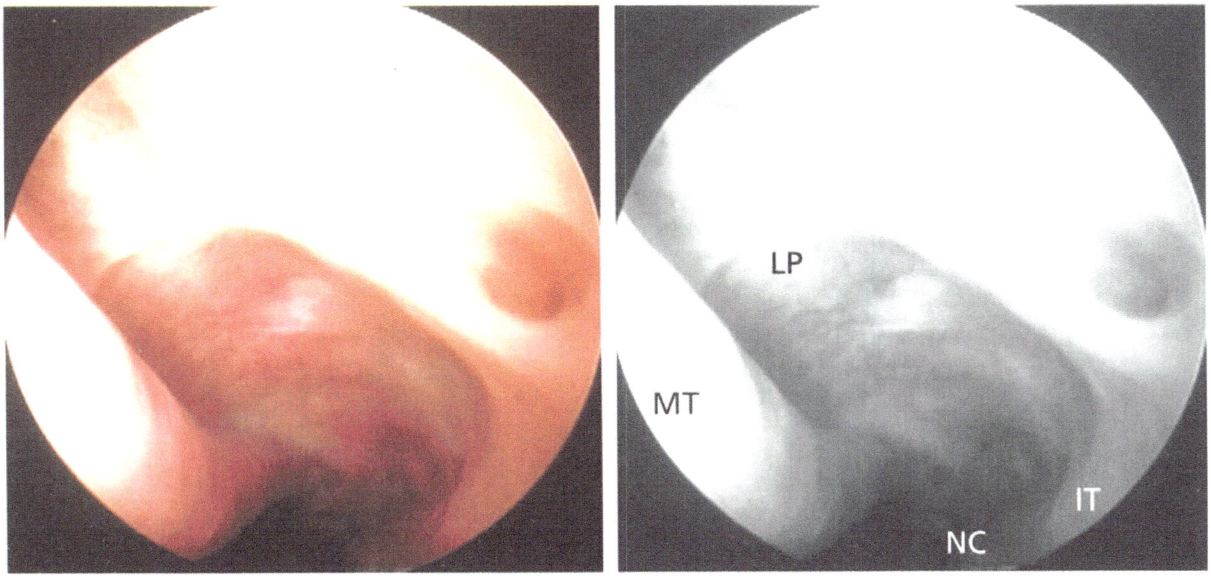

Fig. 12.1. Endoscopic view of middle meatus in Fig. 12.4.

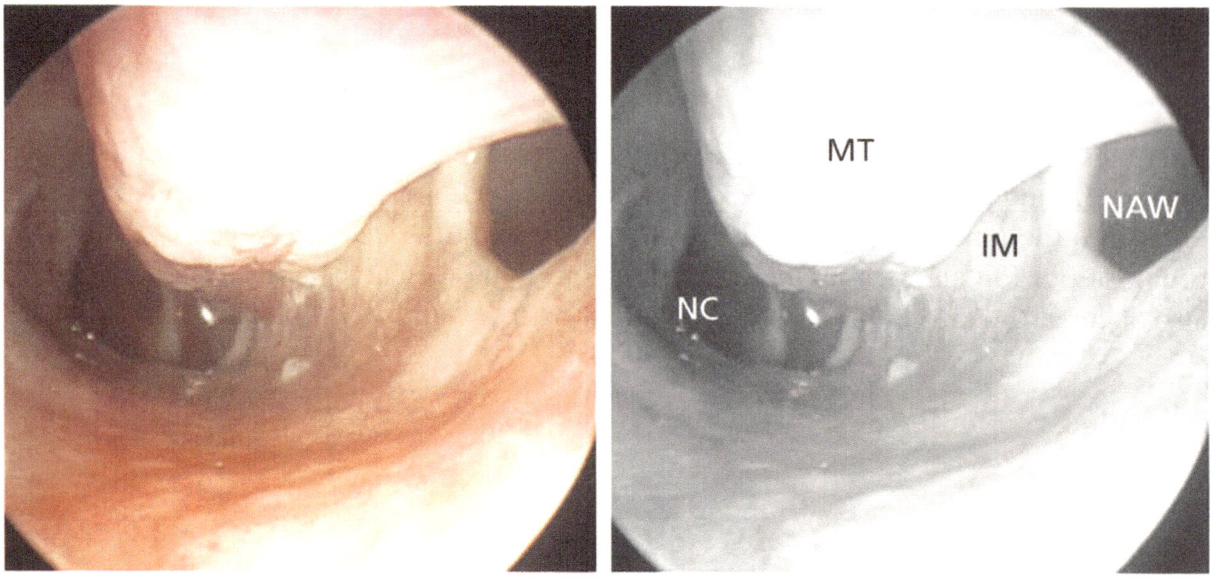

Fig. 12.2. Naso-antral window of maxillary sinus and middle meatus.

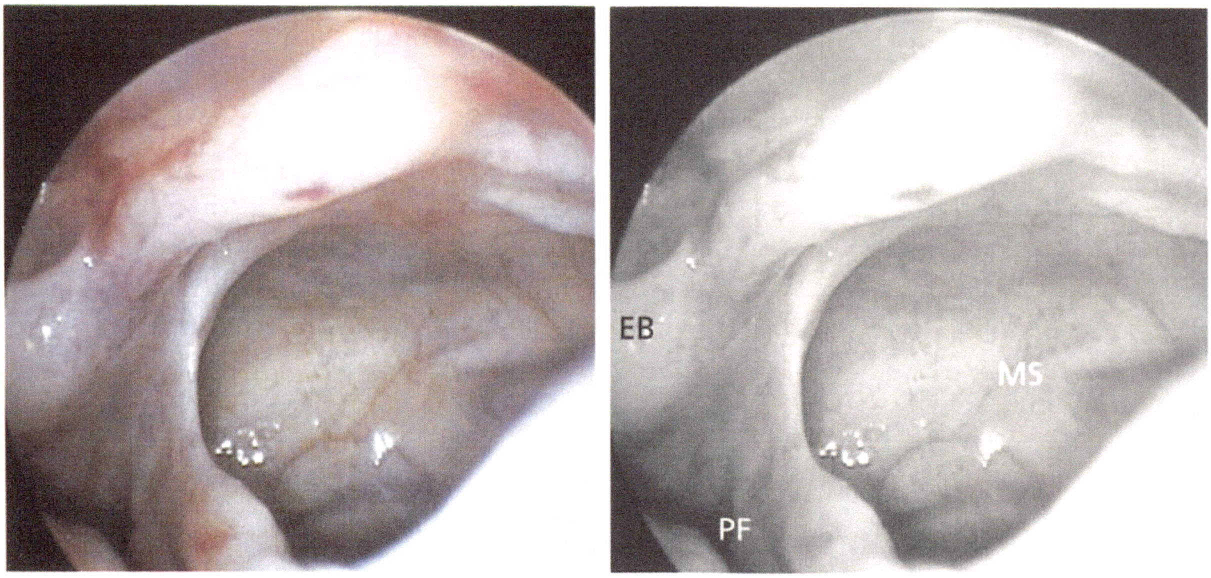

Fig. 12.3. Post-operative middle meatal antrostomy.

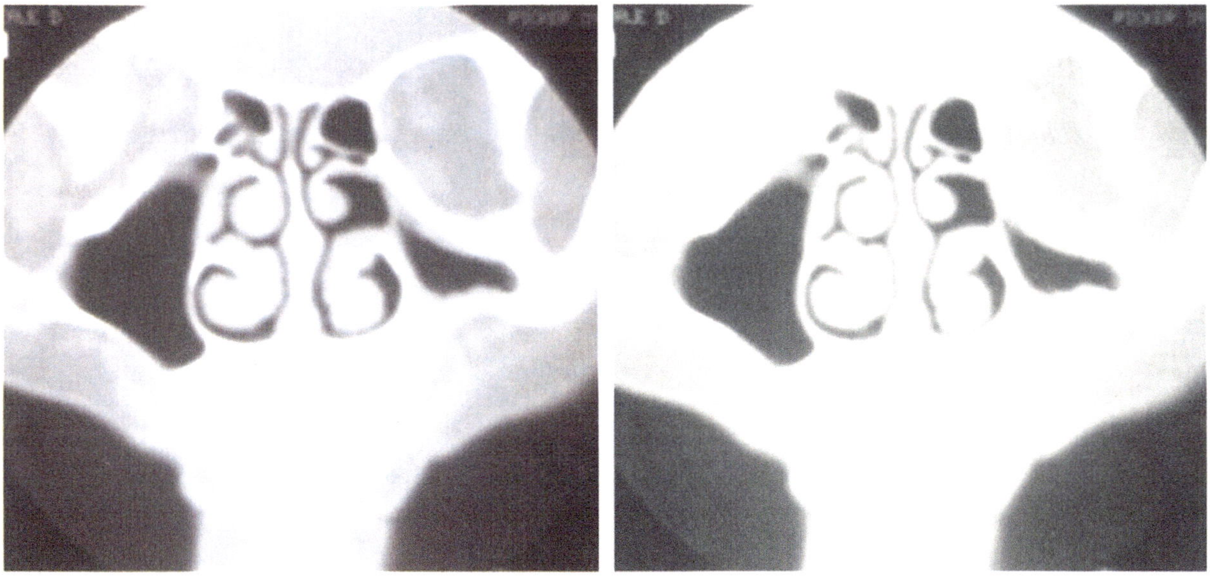

Fig. 12.4. Hypoplastic left maxillary sinus.

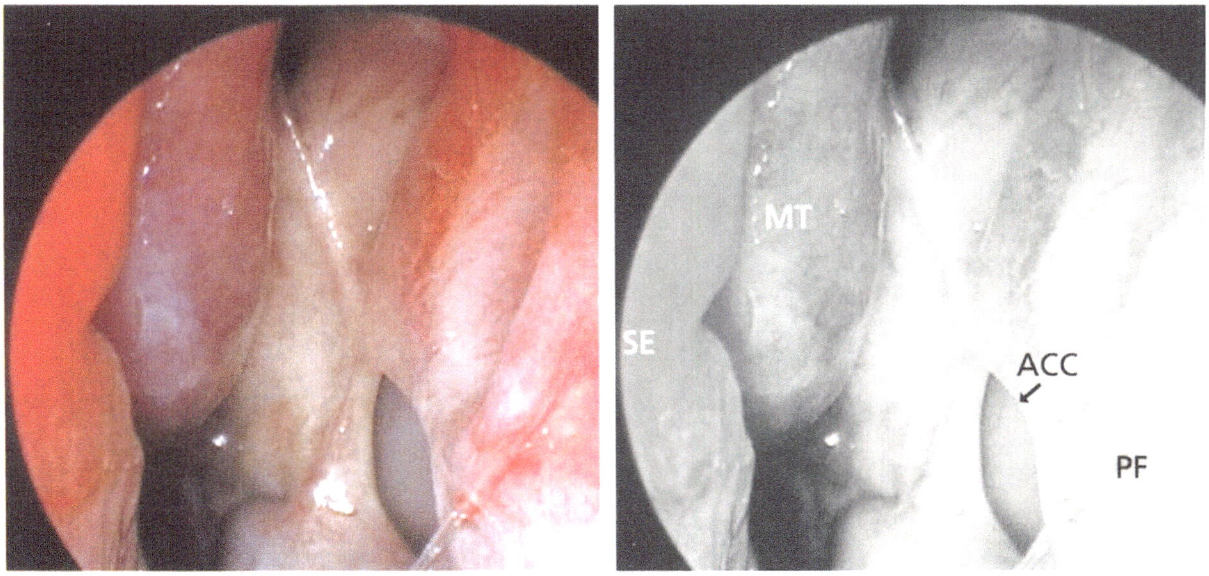

Fig. 12.5. Accessory ostium of maxillary sinus in posterior fontanelle.

13. The posterior ethmoid

Although all ethmoid cells are ultimately continuous with the basal lamella, the latter is used as a reference to divide anterior and posterior cells. The ethmoid complex extends to the face of the sphenoid and may pneumatize superior to the sphenoid, creating the Onodi cell. The cells are continuous with the lamina papyracea laterally. Drainage for the posterior ethmoid cells is into the superior meatus. Occasionally the optic nerve is seen in the most posterior lateral aspect of the posterior ethmoid complex.

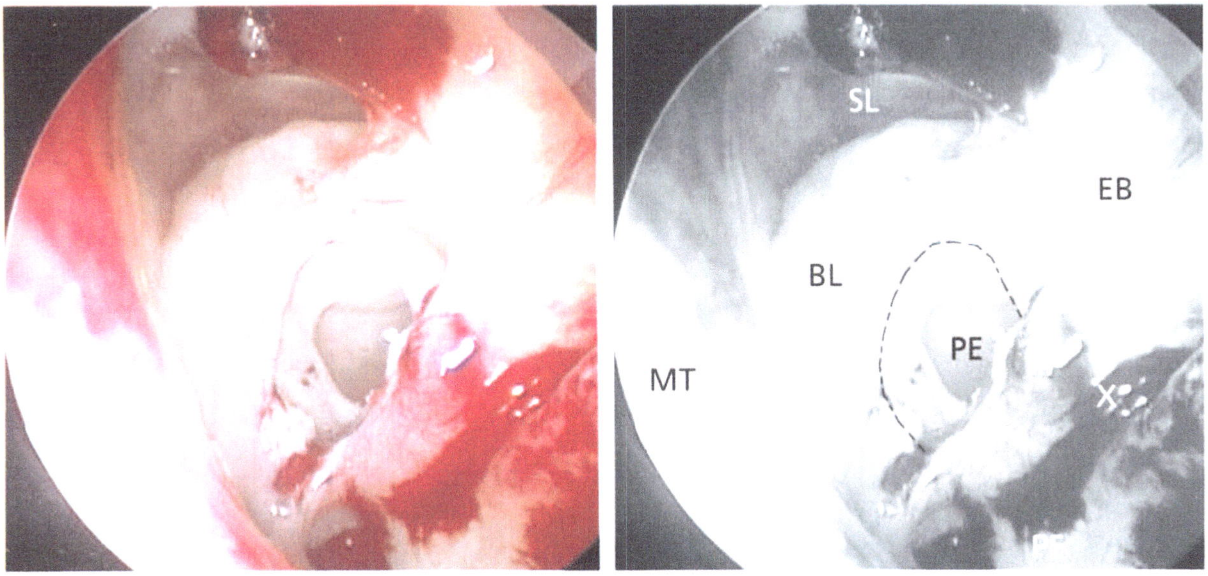

Fig. 13.1. Left side, zero degree lens, basal lamella open, posterior ethmoid visible, base of skull also visible, sinus lateralis above; ethmoidal bulla at 2 o'clock, final common pathway at 3 o'clock. See Fig. 7.7. (p. 62).

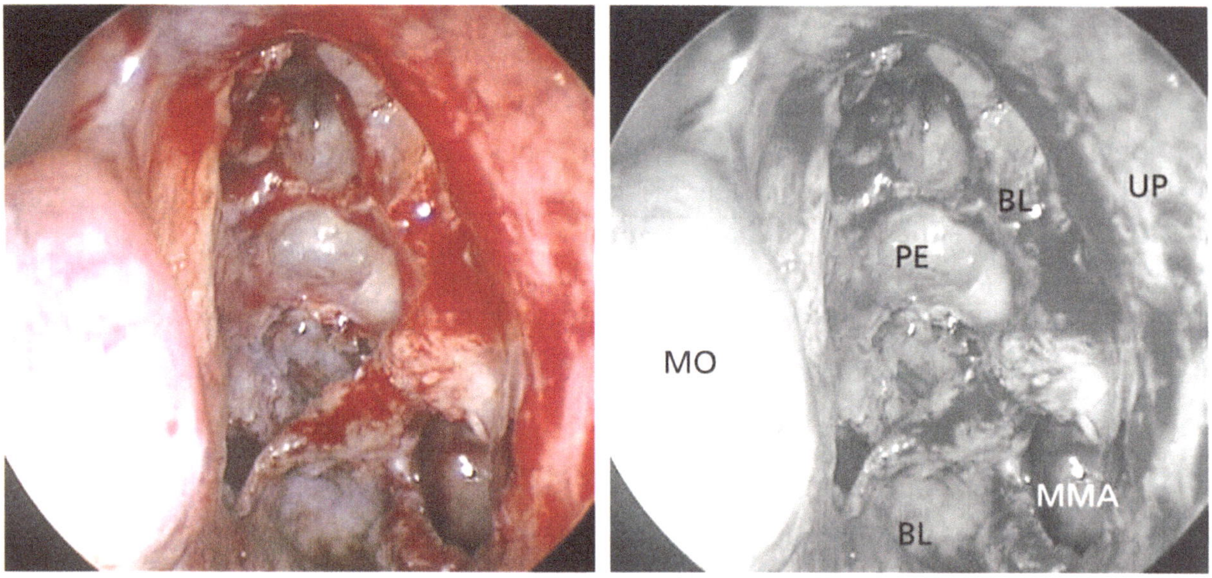

Fig. 13.2. Left side, zero degree lens, complete ethmoid dissection.

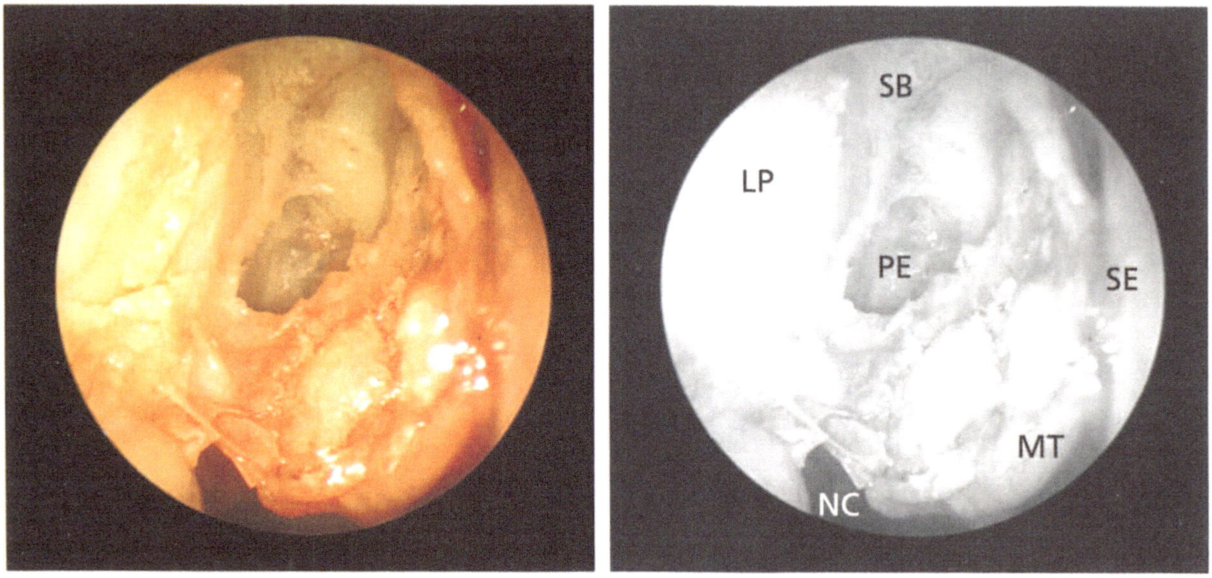

Fig. 13.3. Right side, zero degree lens, anterior and posterior ethmoid complete; base of skull visible.

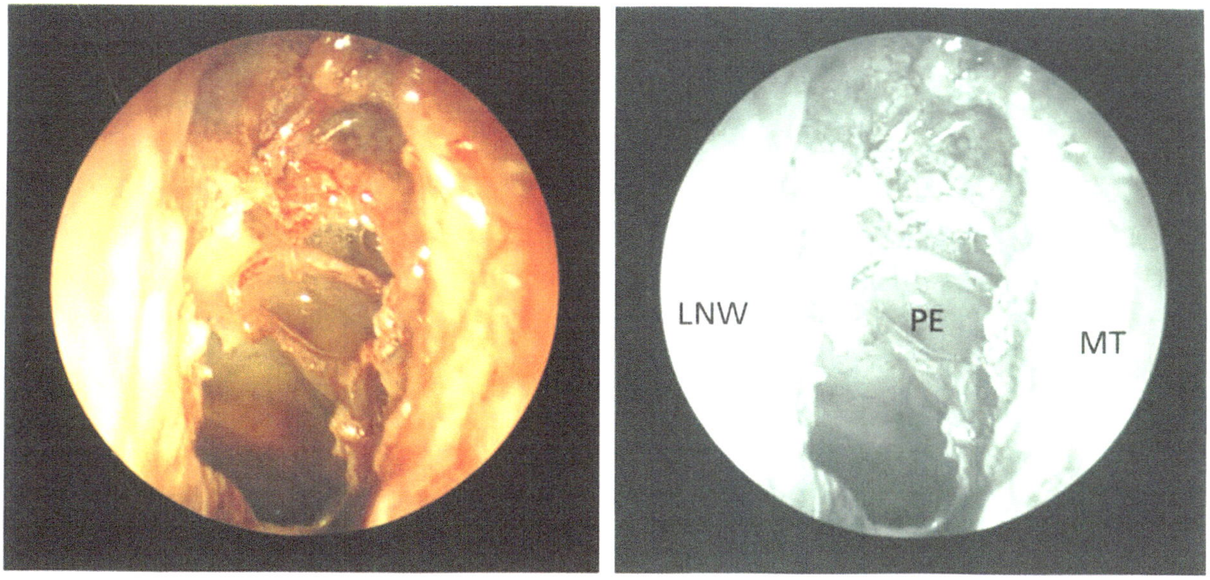

Fig. 13.4. Right side, 30 degree lens, posterior ethmoid dissection to face of sphenoid.

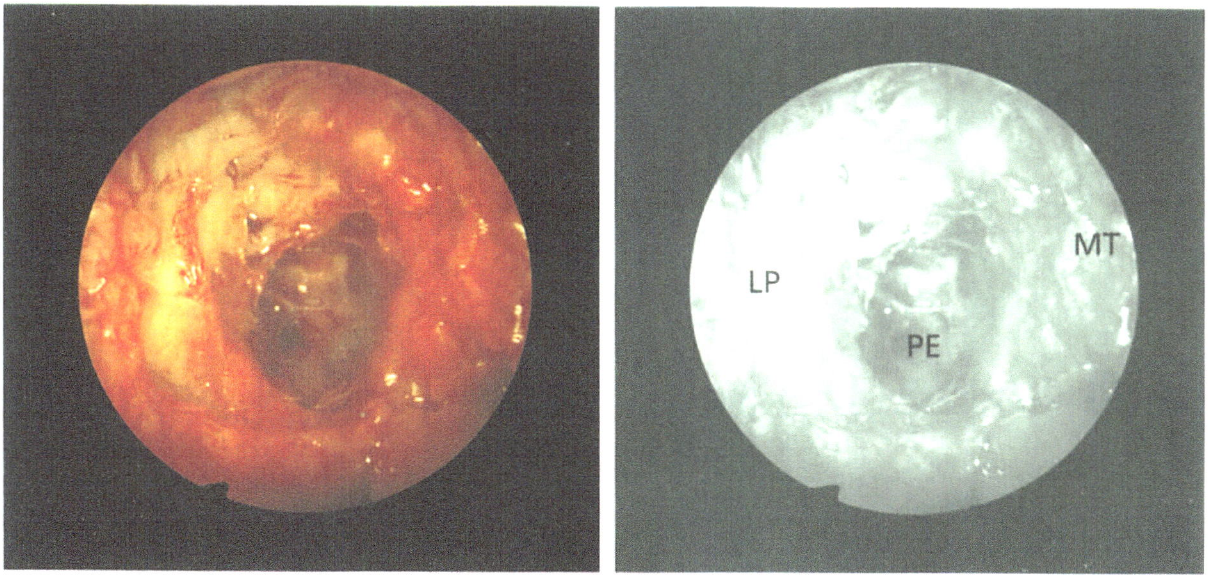

Fig. 13.5. Left side, zero degree lens, posterior ethmoid open.

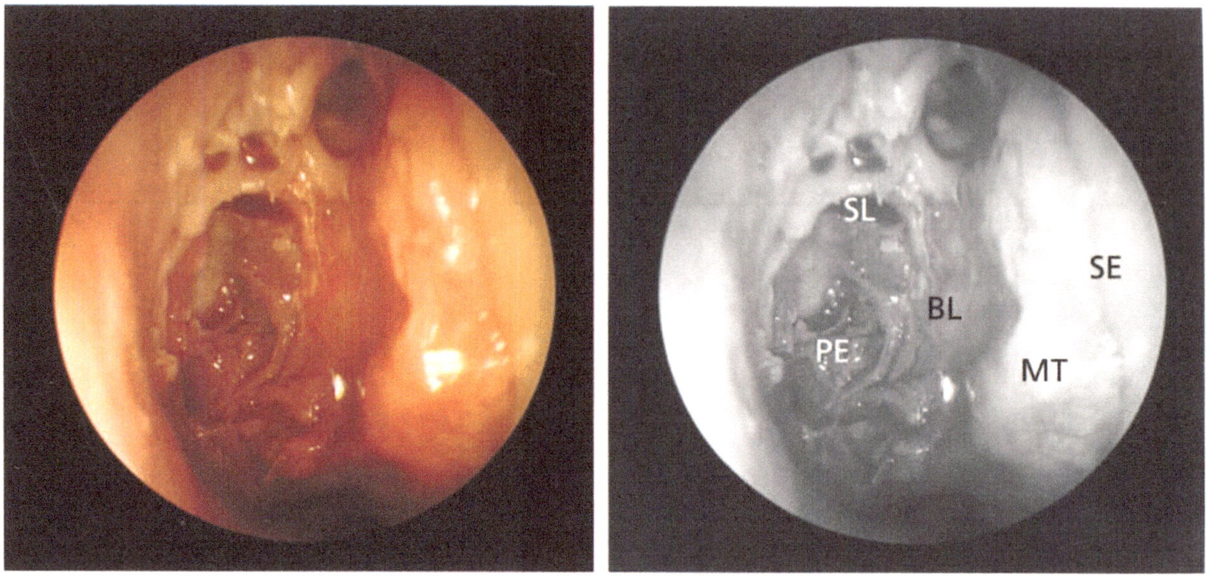

Fig. 13.6. Right side, 30 degree lens, posterior ethmoid and frontal sinus.

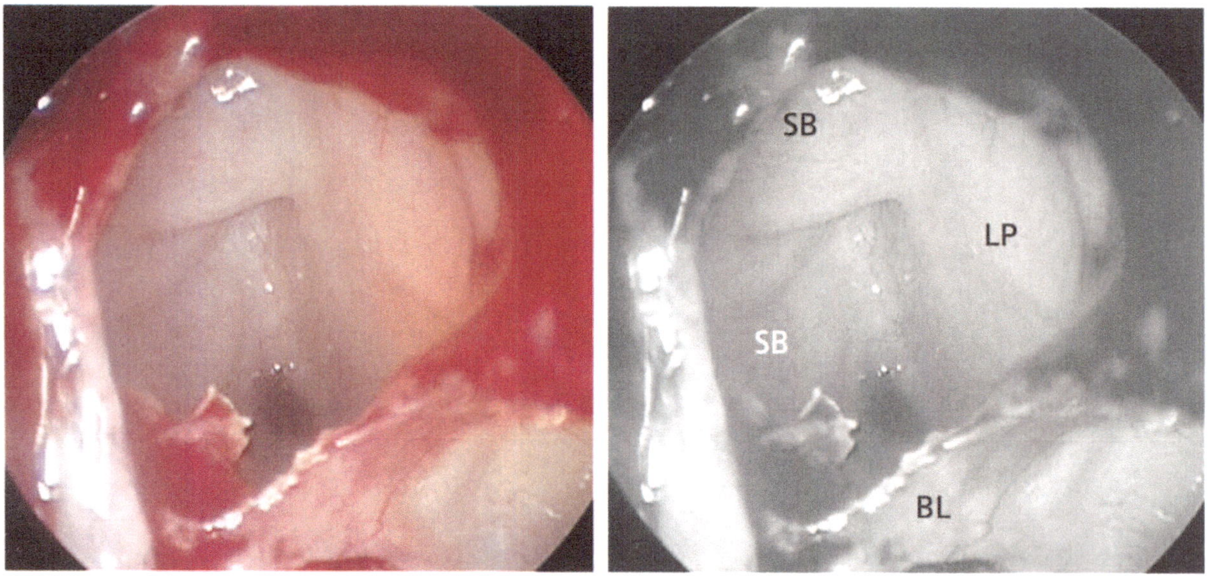

Fig. 13.7. Left side, zero degree lens, extreme posterior ethmoid cell with lamina papyracea and posterior ethmoid and neurovascular bundle.

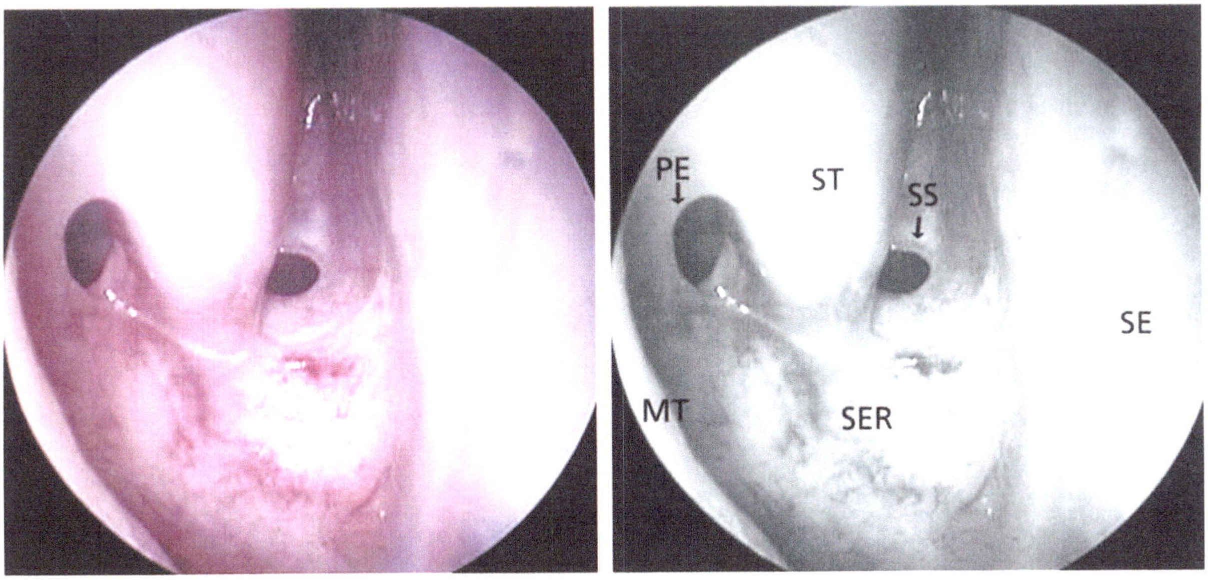

Fig. 13.8. Right side, zero degree lens, ostium of posterior ethmoid and sphenoid sinuses.

14. The superior turbinate

The superior turbinate is continuous with the middle turbinate and the skull base. At endoscopy, it is found between the middle turbinate and the nasal septum. Its projection forms the medial limits of the superior meatus, into which the posterior ethmoid sinuses drain. Tracking from the nasal spine along the posterior one-third of the superior turbinate is a useful method to locate the sphenoid ostium. A concha bullosa may also be present in the superior turbinate.

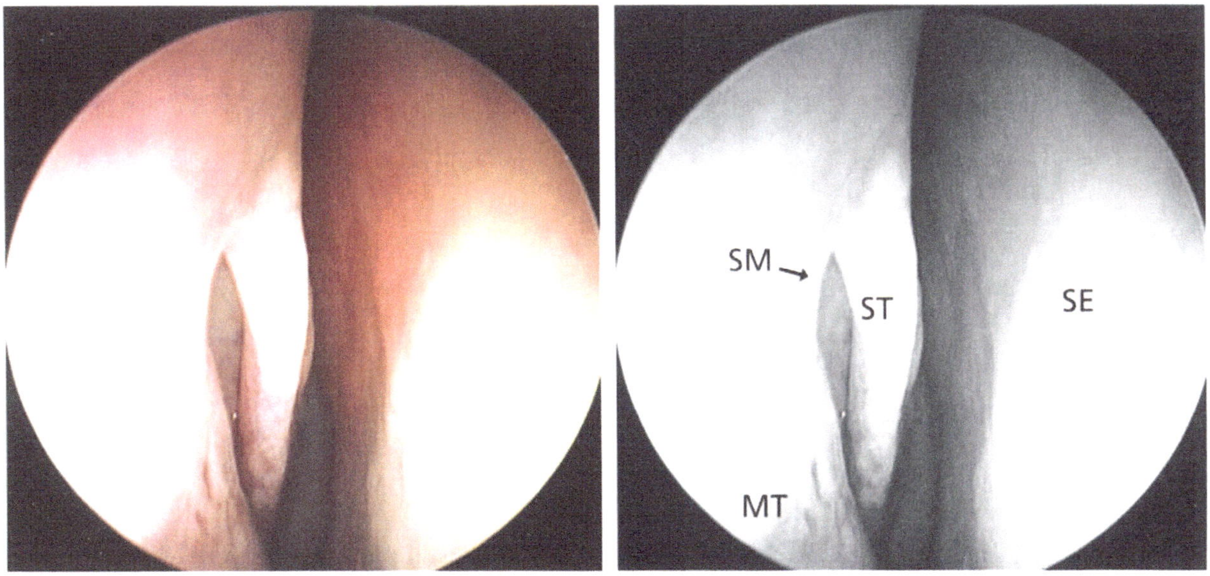

Fig. 14.1. Right side, zero degree lens, superior turbinate and superior meatus.

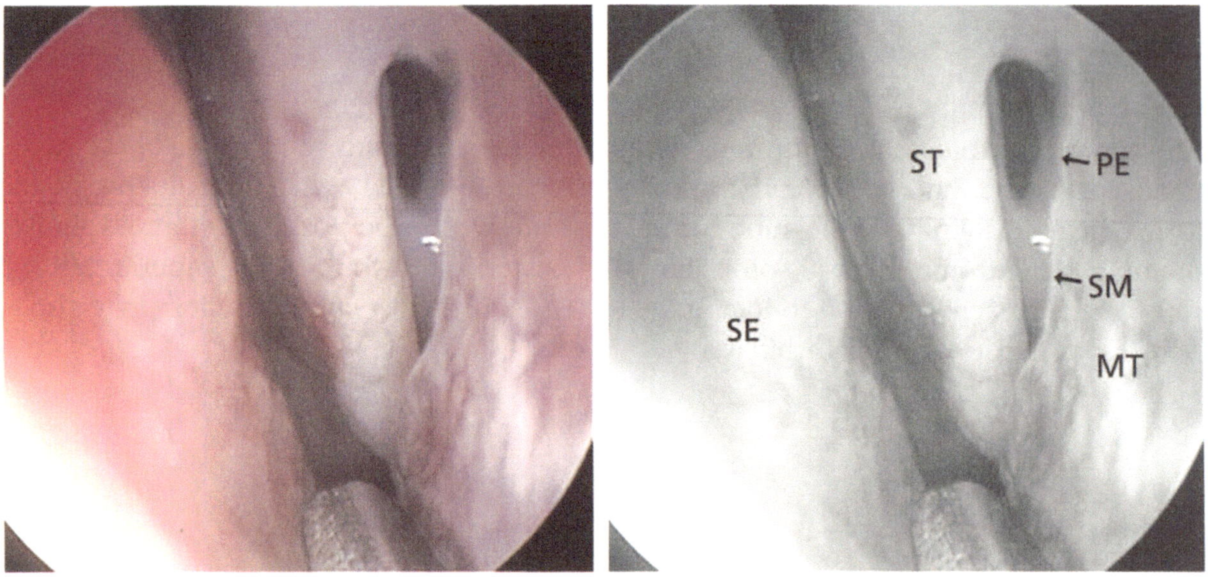

Fig. 14.2. Left side, 30 degree lens, superior turbinate, middle meatus, and ostium of posterior ethmoid sinuses.

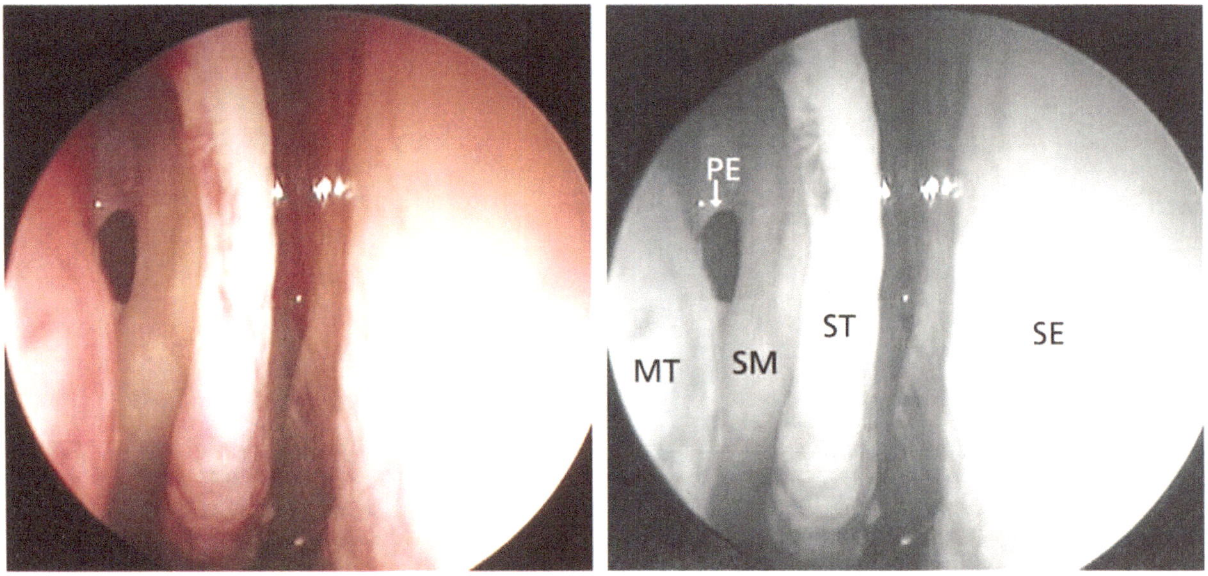

Fig. 14.3. Right side, 30 degree lens, superior turbinate, superior meatus, and ostium of posterior ethmoid cell.

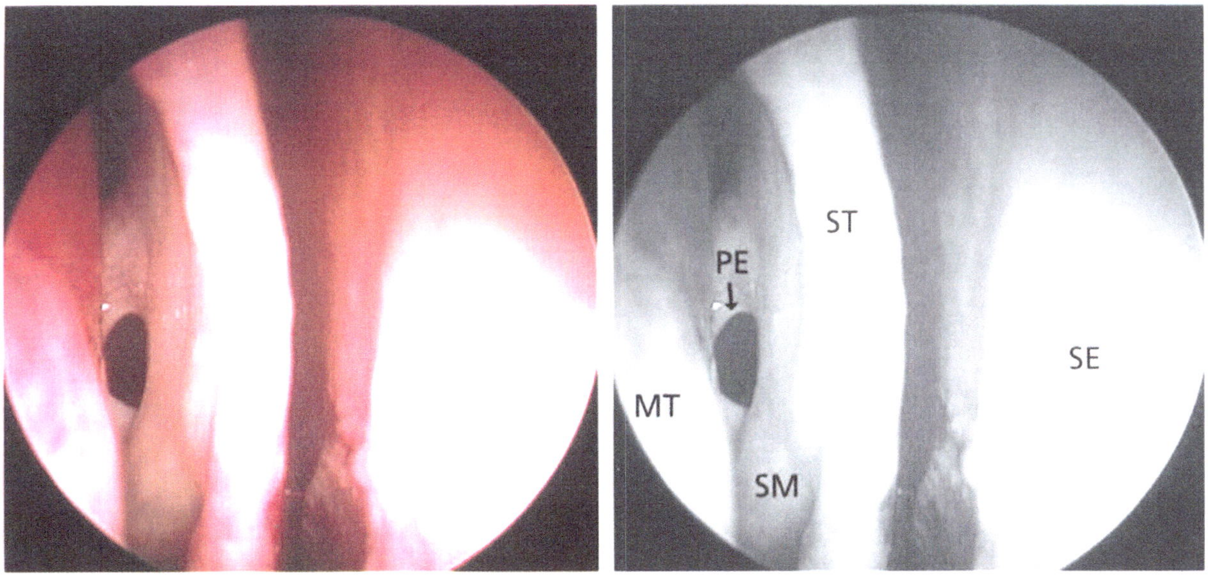

Fig. 14.4. Right side, 30 degree lens, superior turbinate, superior meatus, and ostium of posterior ethmoid cells.

15. The sphenoid sinus

In most cases, the sphenoid sinus is the most posterior of the paranasal sinuses. It drains directly into the nasal cavity by an ostium in its anterior-superior medial aspect, emptying into the sphenoethmoid recess. It is located by the endoscopist by following the line from the nasal spine to the superior meatus and tracking along the posterior one-third of the superior turbinate. The ostium is variable in size and in its relationship to both the superior turbinate and the sphenoethmoid recess.

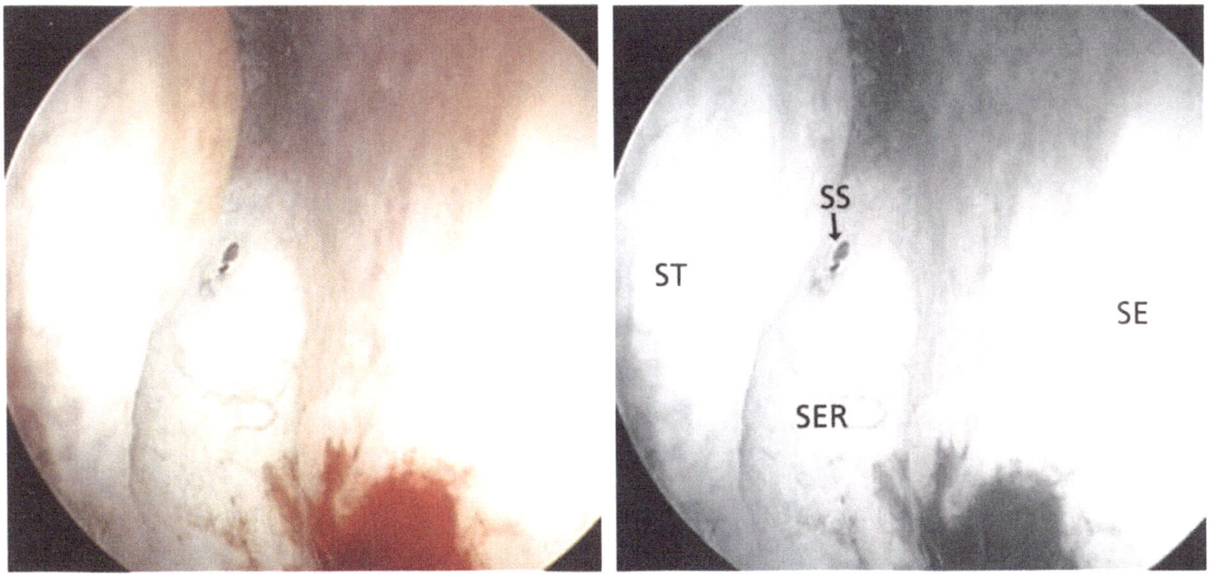

Fig. 15.1. Right side, zero degree lens, very small sphenoid ostium in quiescent sphenoid sinus. Superior turbinate is to the left of the picture.

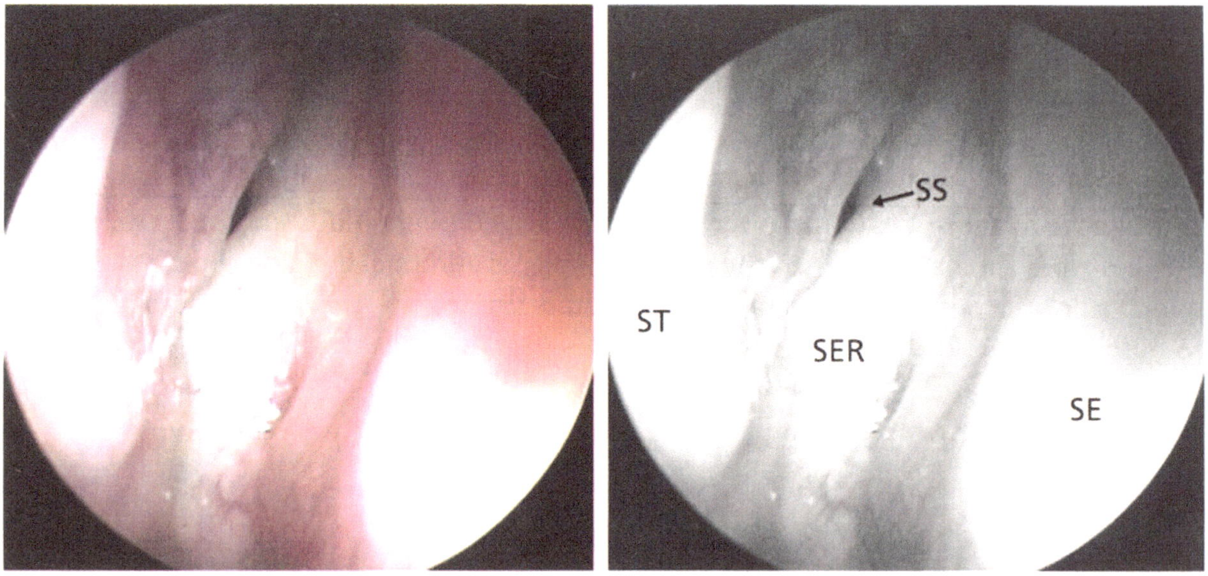

Fig. 15.2. Right side, zero degree lens, superior turbinate has been pressed laterally to partially expose a larger sphenoid ostium in the sphenoethmoid recess.

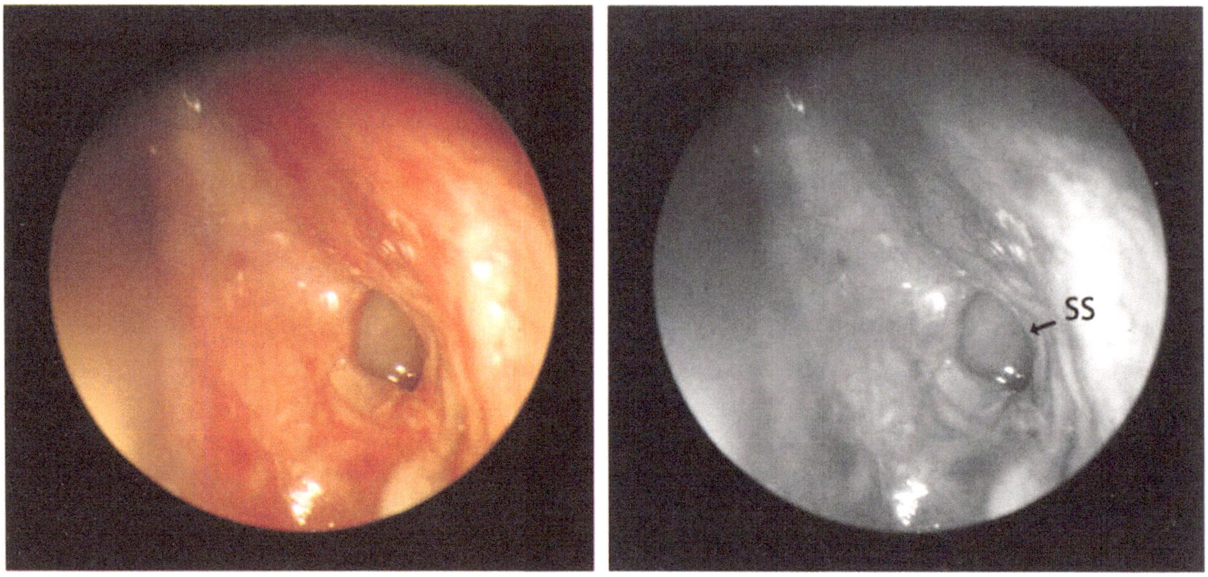

Fig. 15.3. Left side, zero degree lens, sphenoid ostium.

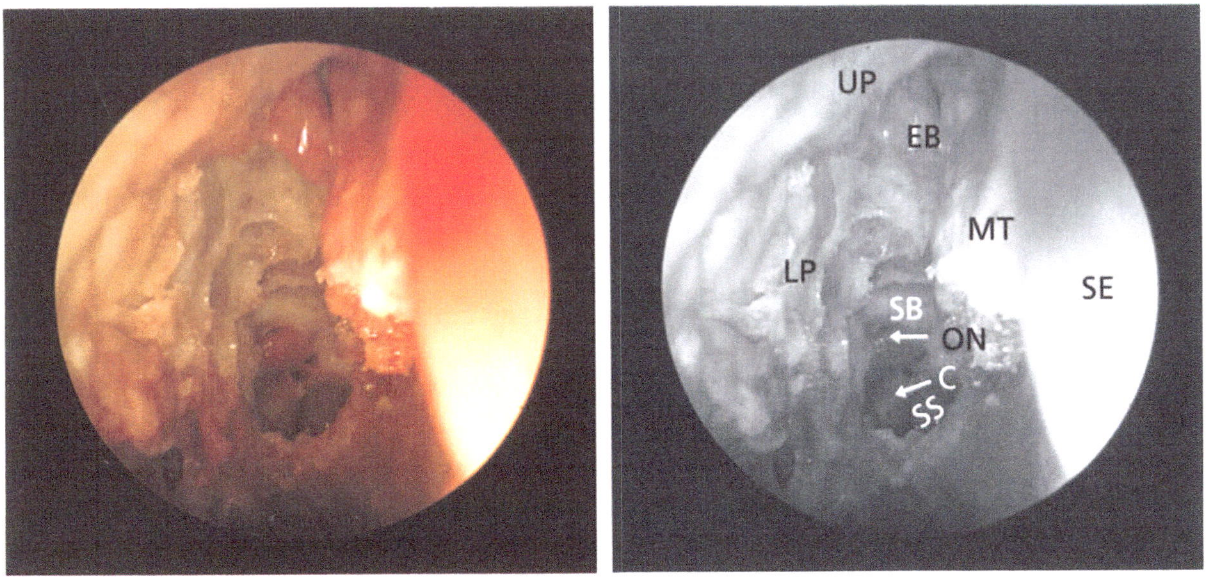

Fig. 15.4. Right side, zero degree lens, optic nerve visible in sphenoid sinus after removal of overlying bone.

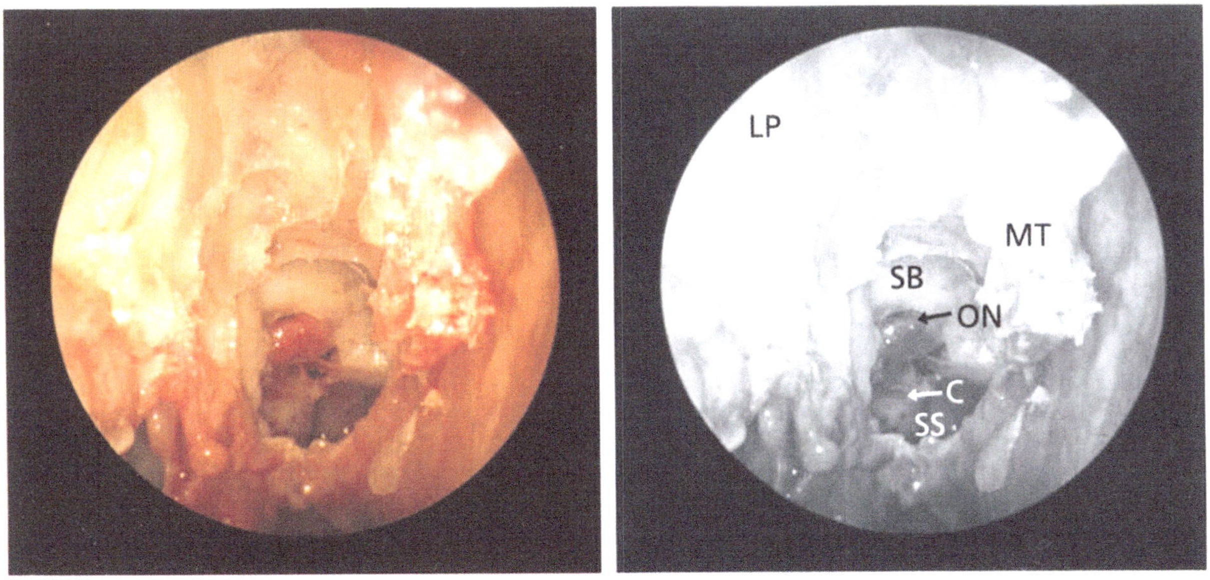

Fig. 15.5. Right side, zero degree lens, close-up of Fig. 15.4. Ribbed vault at 12 o'clock.

16. The skull base (ribbed vault)

The anterior skull base is the roof of the ethmoid complex. It is thinner and therefore more vulnerable anteriorly than posteriorly. Both anterior and posterior ethmoid neurovascular bundles may be seen if a complete marsupialization of the ethmoid complex is done.

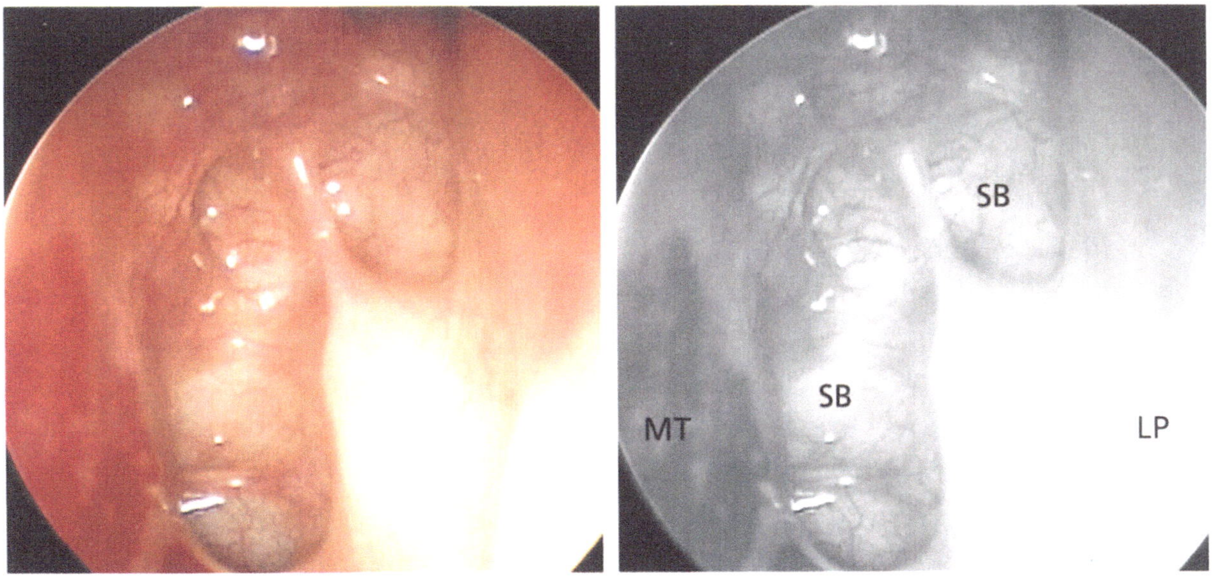

Fig. 16.1. Right side, 30 degree lens, skull base, post-operative skull base showing ribbed vault and neurovascular bundle.

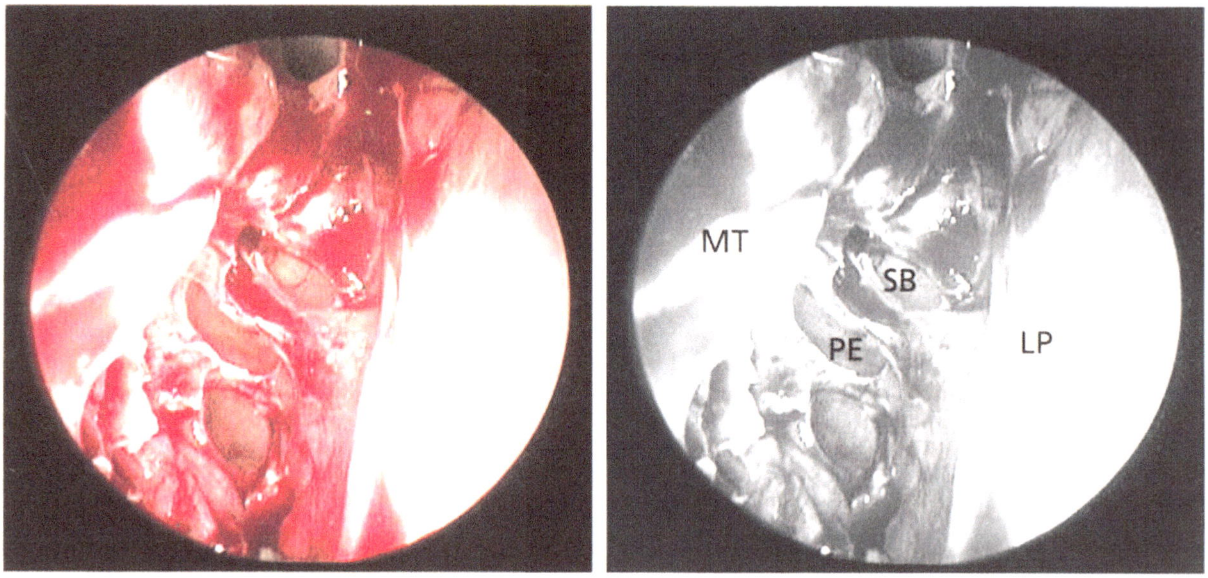

Fig. 16.2. Right side, 30 degree lens, lamina papyracea, posterior ethmoid, middle turbinate, left to right; posterior ethmoid cells and skull base following dissection.

17. The frontal sinus

The frontal bone may be extensively pneumatized or not at all. Although any cells above the orbit may be called "frontal", ethmoid development into the frontal bone is a common occurrence.

The frontal sinus is divided by a partition. It is the neighbor of the forehead, the orbit, and the endocranium.

Problems associated with the frontal sinus are usually related to drainage through a space that occurs as a duct or an infundibulum and that may be distorted. The literature about this topic is very extensive.

A critical anatomic landmark for the endoscopist is the agger nasi cell. In general, the frontal sinus drains posteriorly and medially to the agger nasi cell, the transition space between the ostium and entrance into the nasal cavity varying in dimensions. The size of the agger nasi cell determines the dimensions of the space available for drainage.

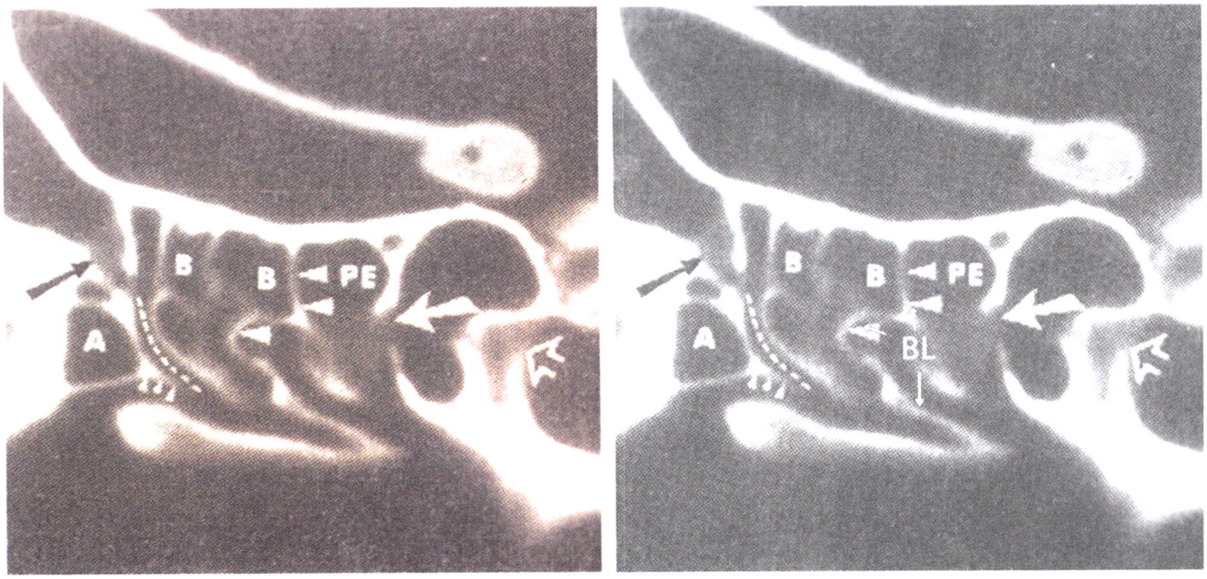

Fig. 17.1. Left side, zero degree lens, a near parasagittal cut through the sphenoethmoid complex. Note the drainage of the frontal sinus posterior and therefore medial to the agger nasi cell. Horizontal and vertical portions of basal lamella are also visible.

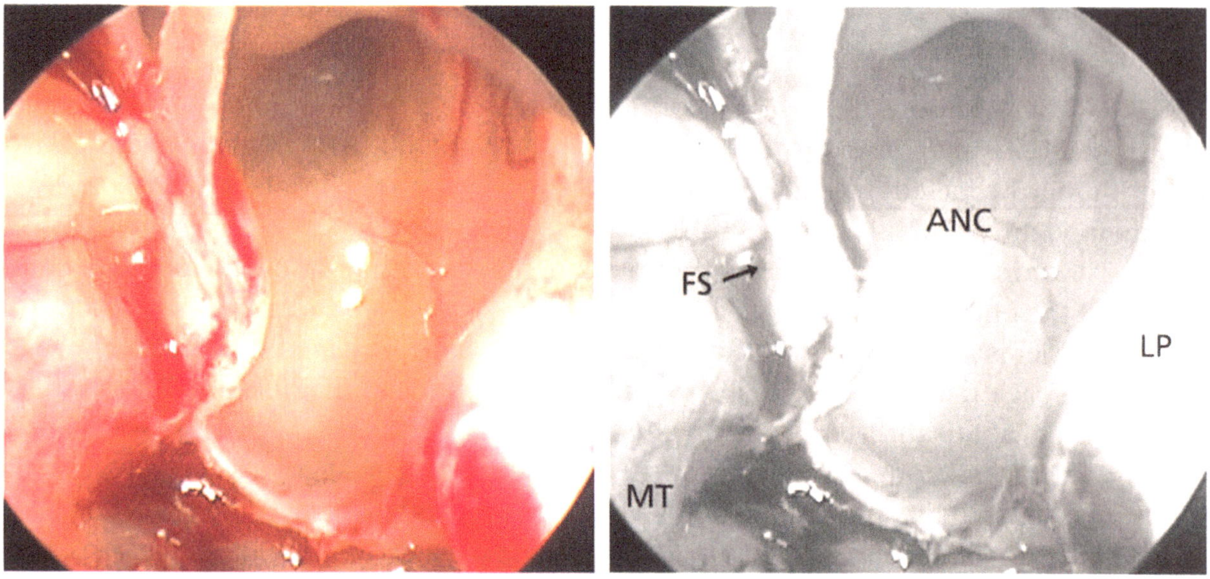

Fig. 17.2. Left side, 30 degree lens, large agger nasi cell with transition space for frontal sinus exposed posteriorly and medially.

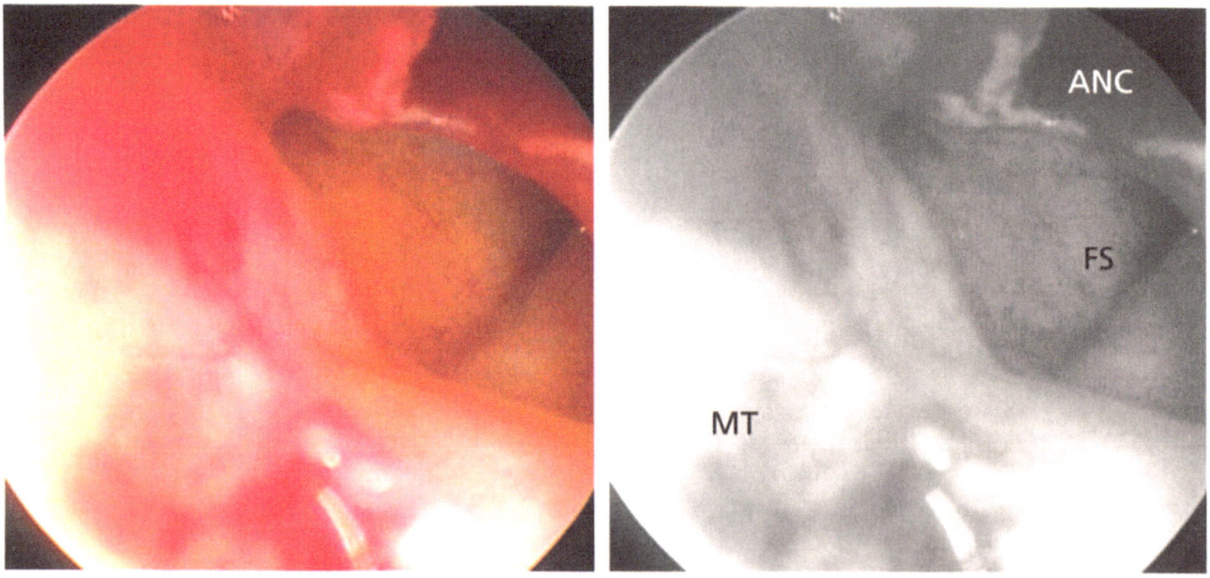

Fig. 17.3. Left side, 30 degree lens, same patient after removal of posterior and medial wall of agger nasi cell. View of frontal sinus ostium.

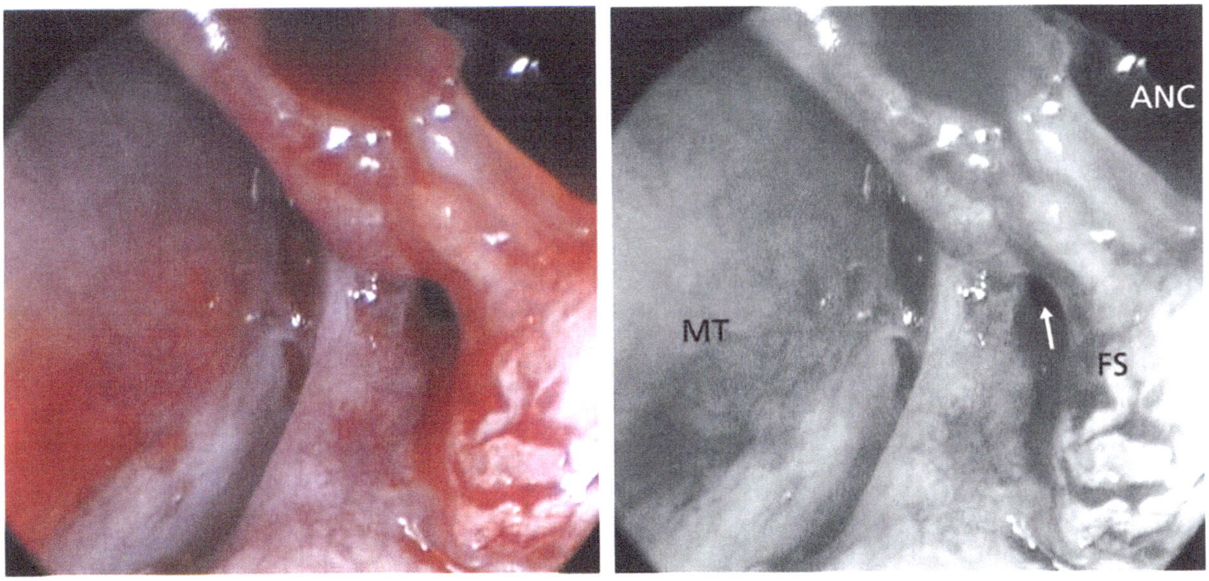

Fig. 17.4. Left side, 30 degree lens, frontal sinus ostium posterior and medial to agger nasi cell.

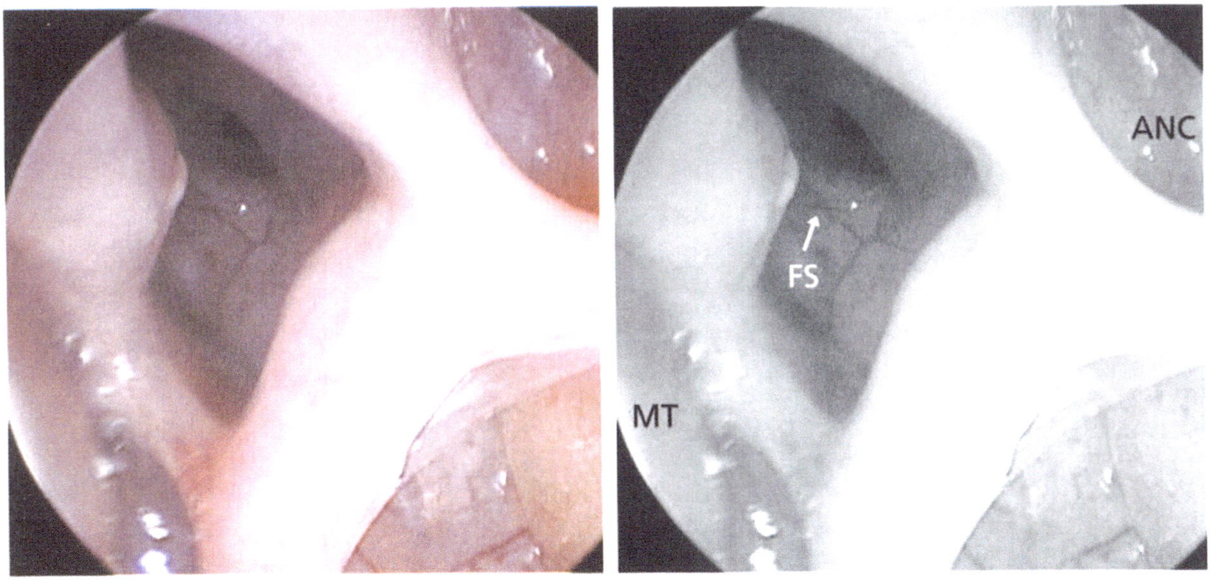

Fig. 17.5. Left side, 30 degree lens, a wide transition space and frontal sinus ostium posterior and medial to agger nasi cell; anterior ethmoid neurovascular bundle seen below the ostium.

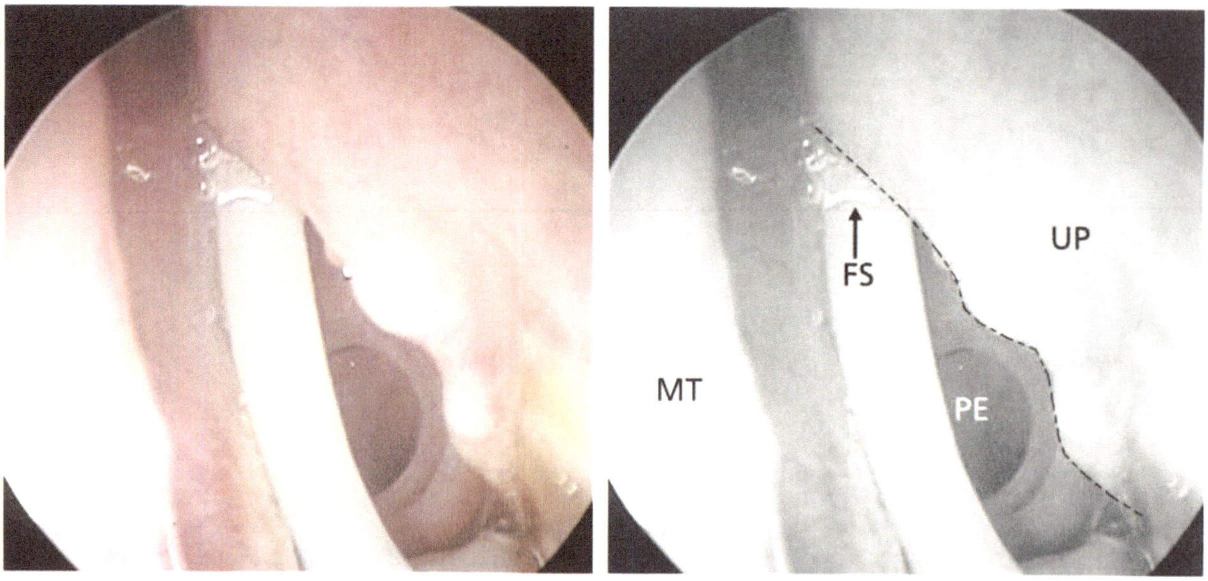

Fig. 17.6. Left side, 30 degree lens, residual uncinate with drainage from frontal sinus.

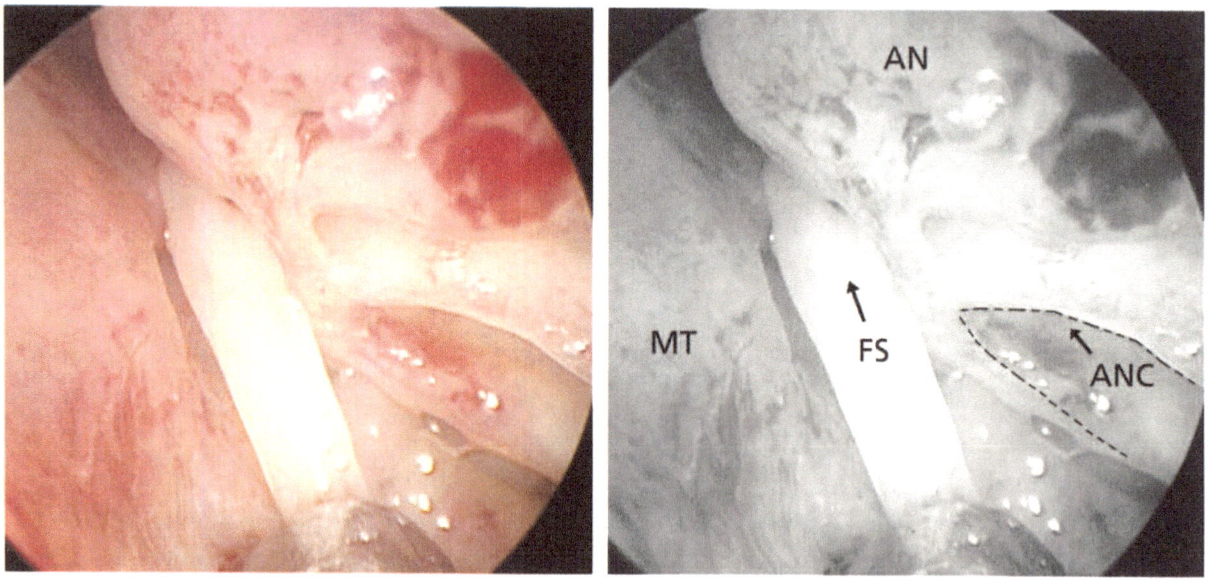

Fig. 17.7. Left side, 30 degree lens, agger nasi cell partially open; secretions from frontal sinus.

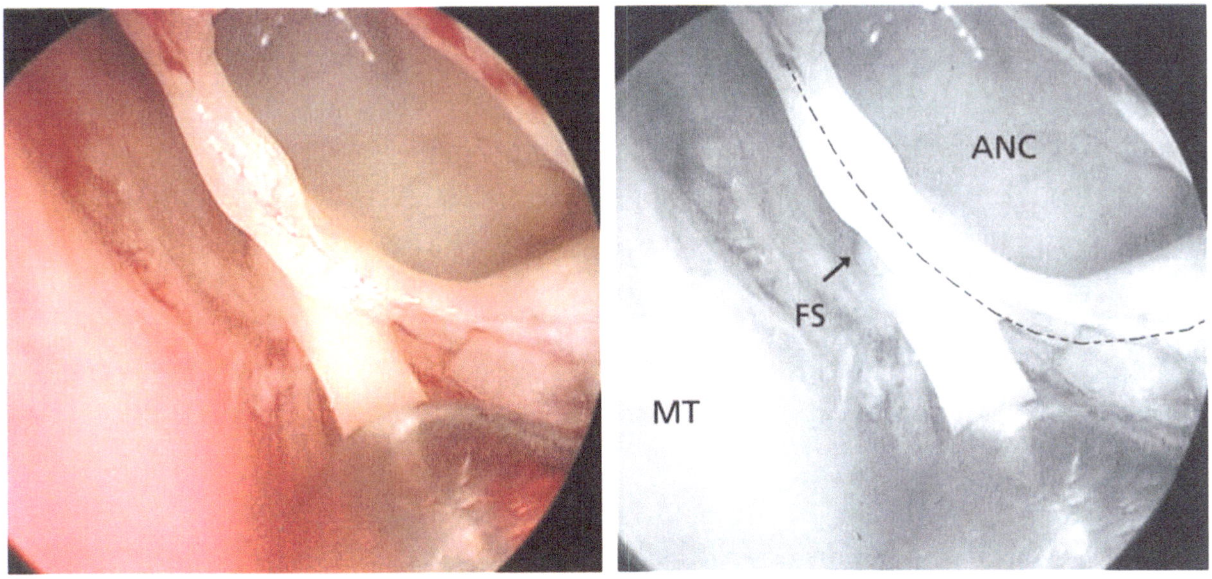

Fig. 17.8. Left side, 30 degree lens, same patient, partial removal of posterior and medial wall; secretions from frontal sinus.

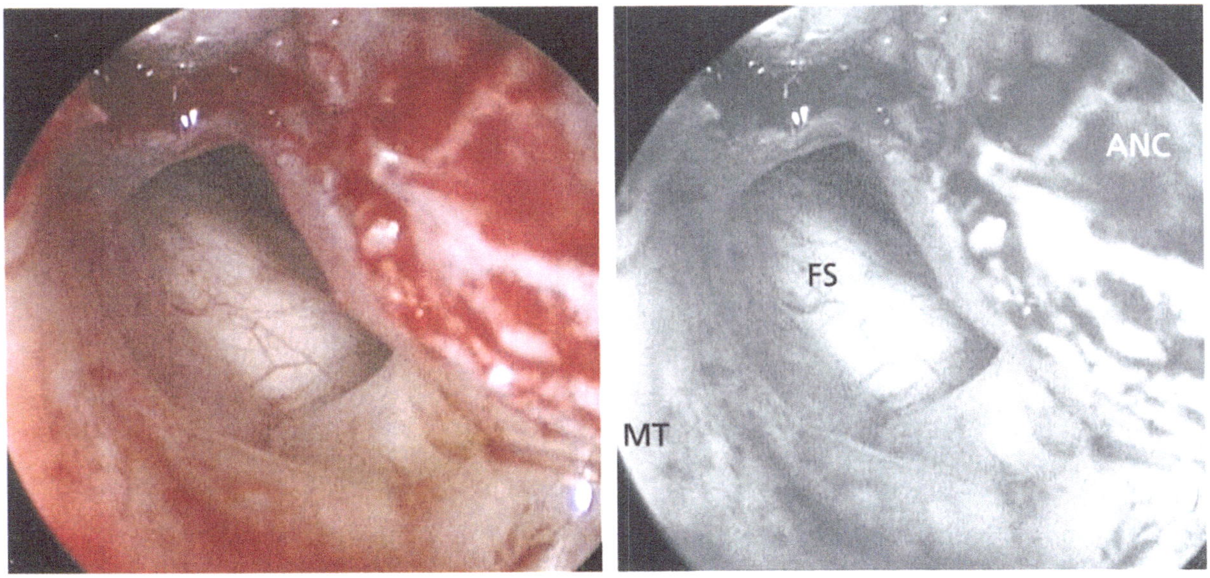

Fig. 17.9. Left side, zero degree lens, same patient, entrance to frontal sinus after posterior and medial wall of agger nasi cell removed; secretions from frontal sinus.

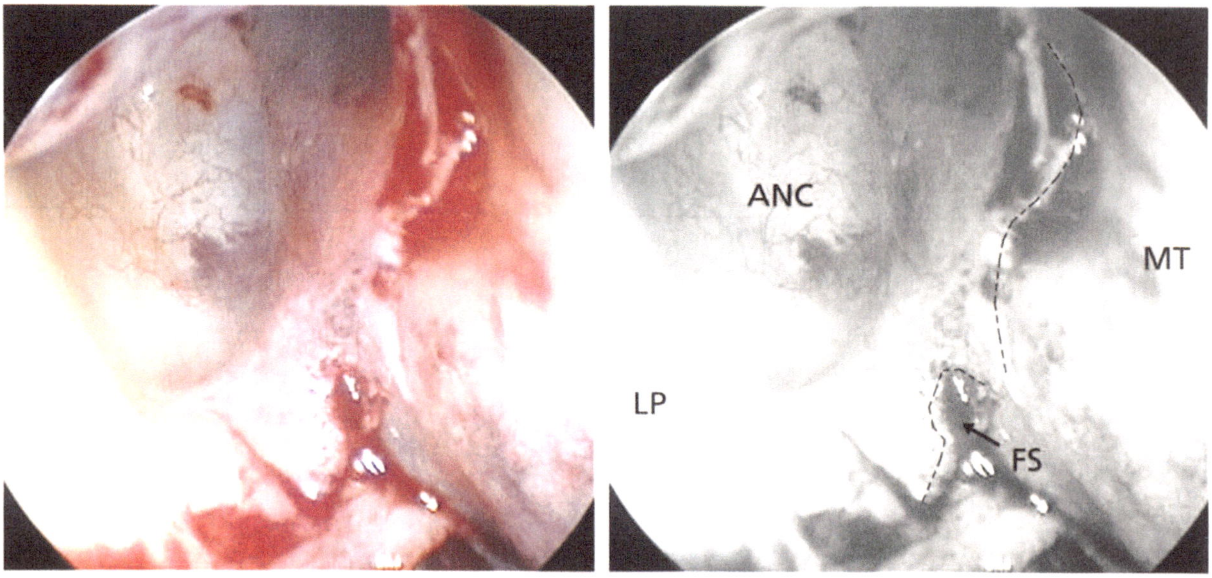

Fig. 17.10. Right side, zero degree lens, large agger nasi with no obvious frontal openings.

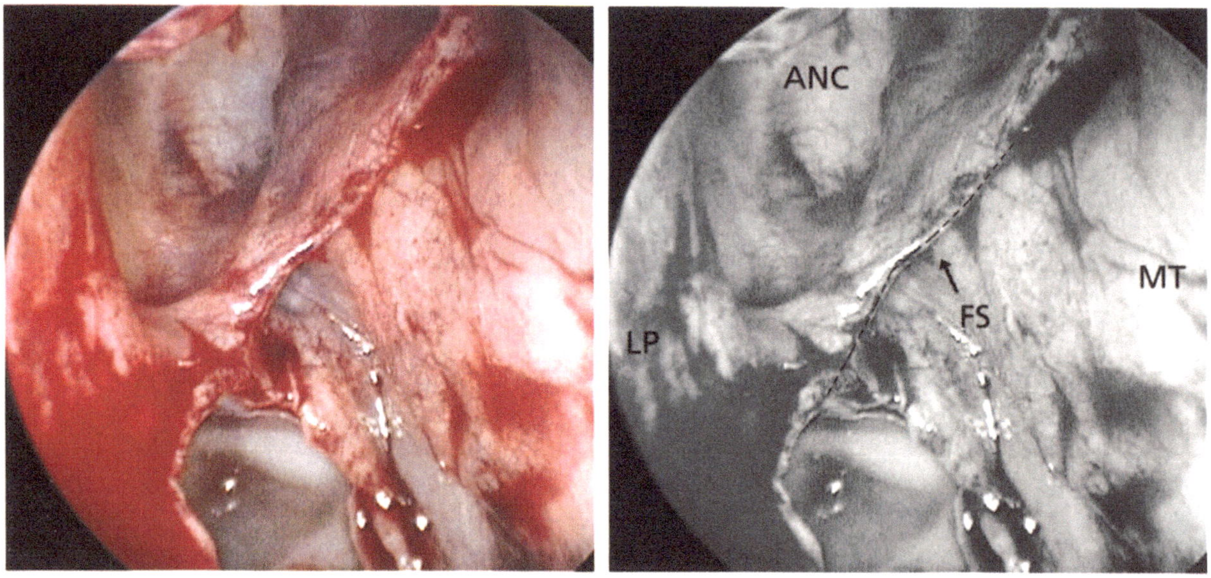

Fig. 17.11. Right side, 30 degree lens, same patient, after minimal removal of posterior and medial wall of agger nasi.

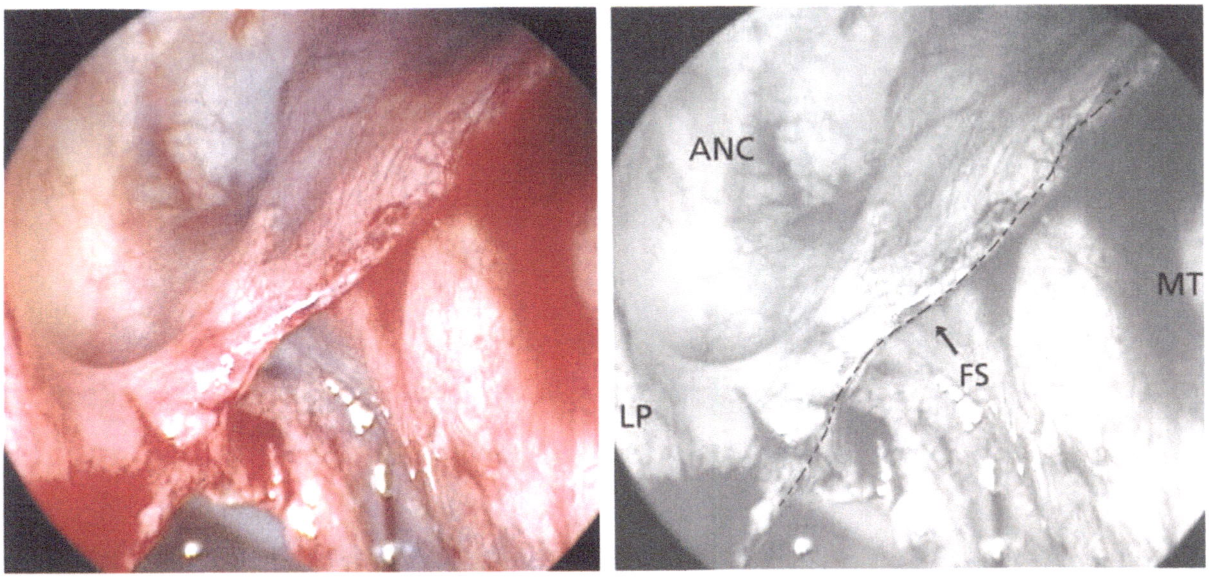

Fig. 17.12. Right side, 30 degree lens, same patient, close-up.

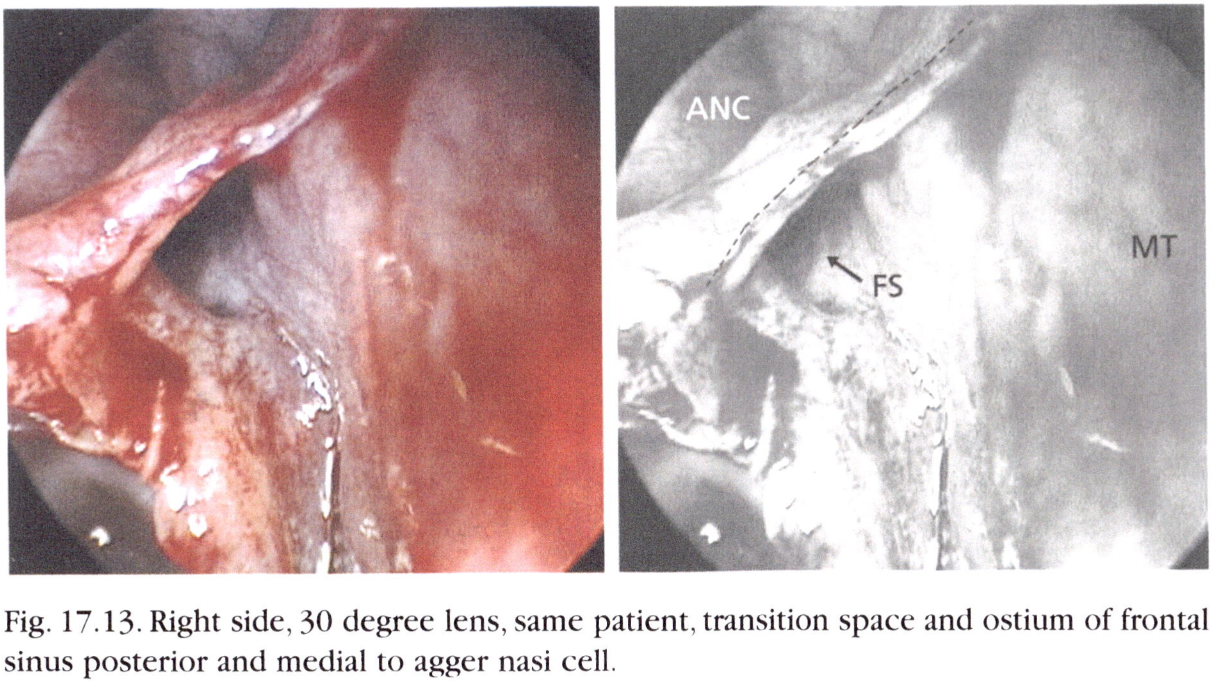

Fig. 17.13. Right side, 30 degree lens, same patient, transition space and ostium of frontal sinus posterior and medial to agger nasi cell.

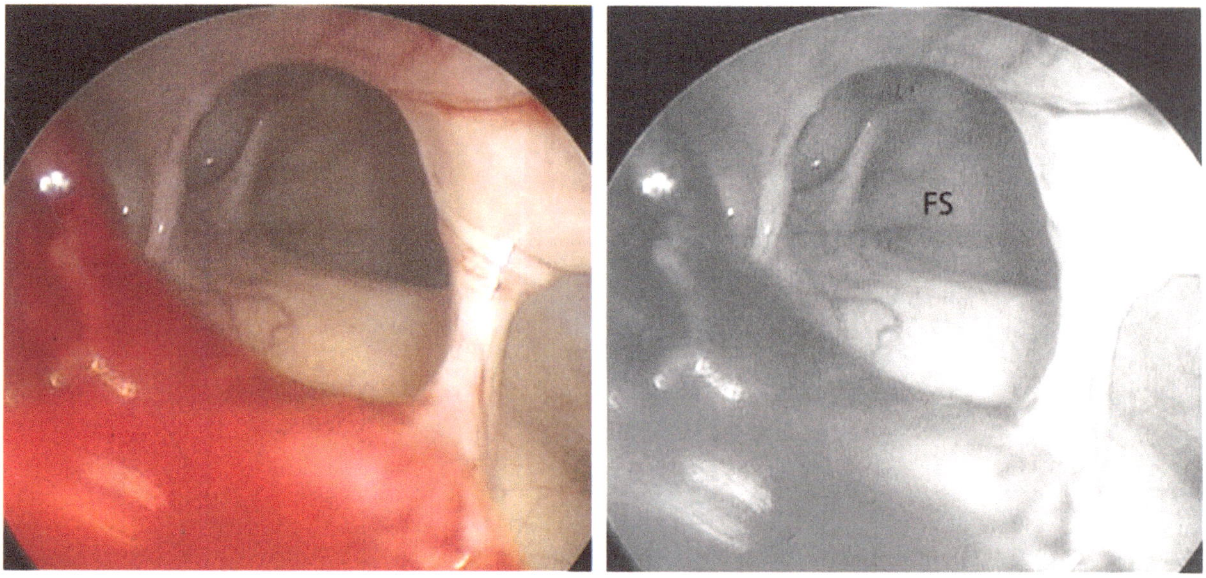

Fig. 17.14. Left side, 30 degree lens, interior of frontal sinus.

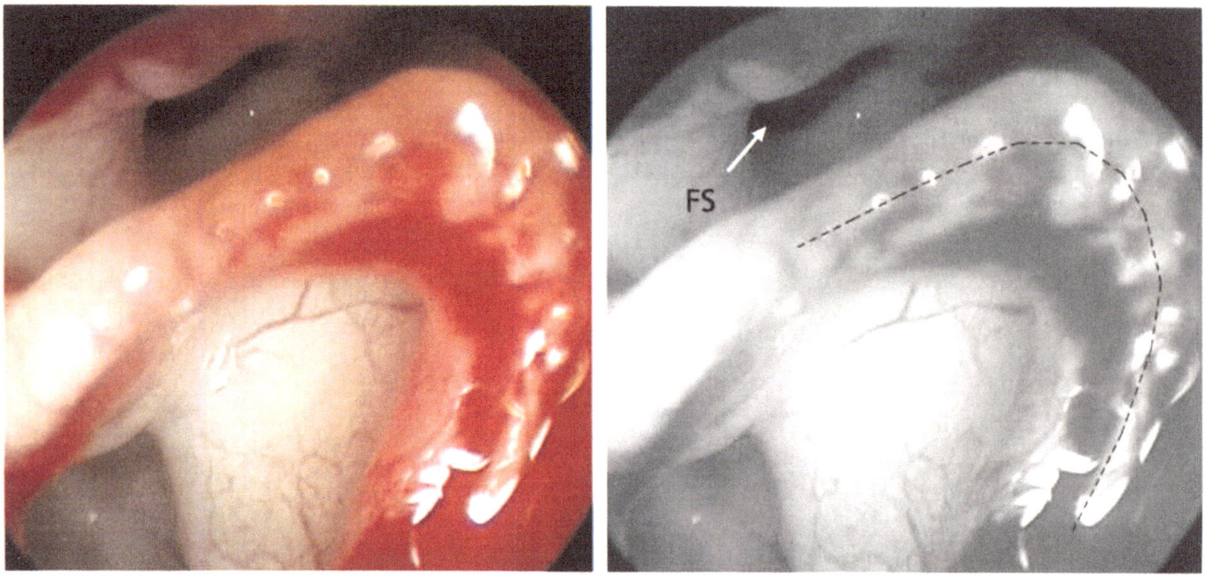

Fig. 17.15. Left side, 30 degree lens, transition space and ostium of frontal sinus.

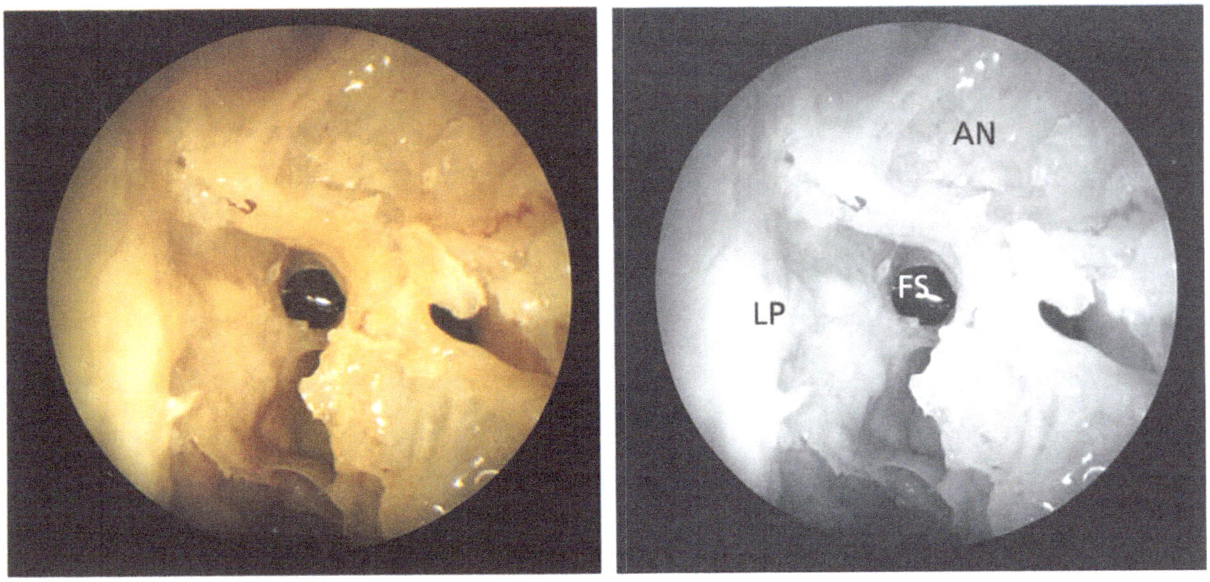

Fig. 17.16. Right side, 30 degree lens, large non-surgical frontal sinus opening.

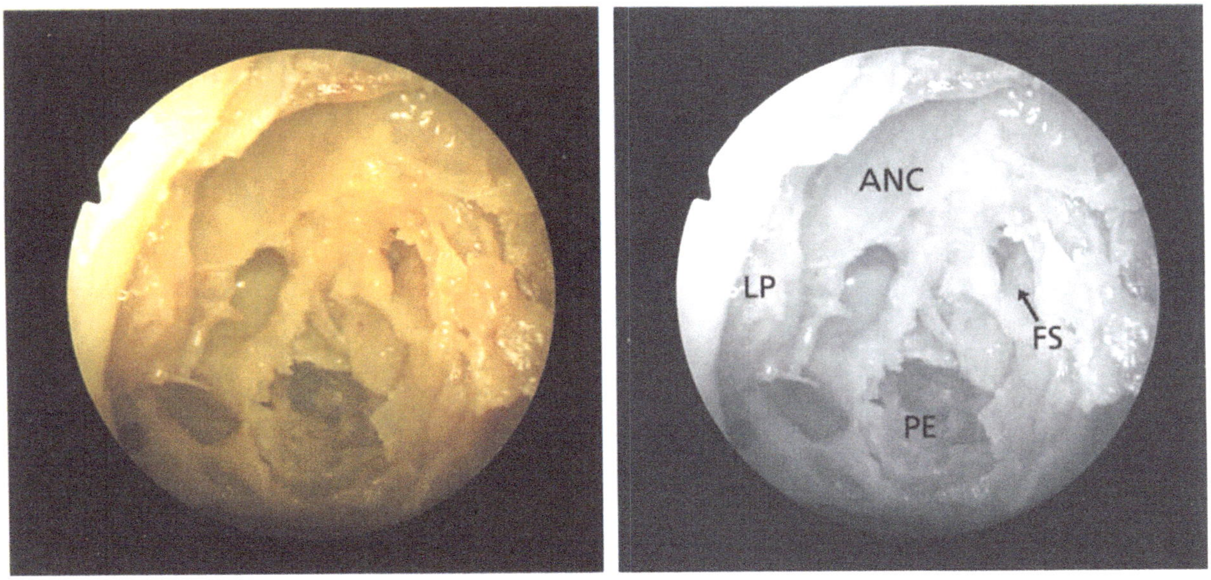

Fig. 17.17. Right side, 30 degree lens, frontal drainage posterior and medial to the agger nasi cell.

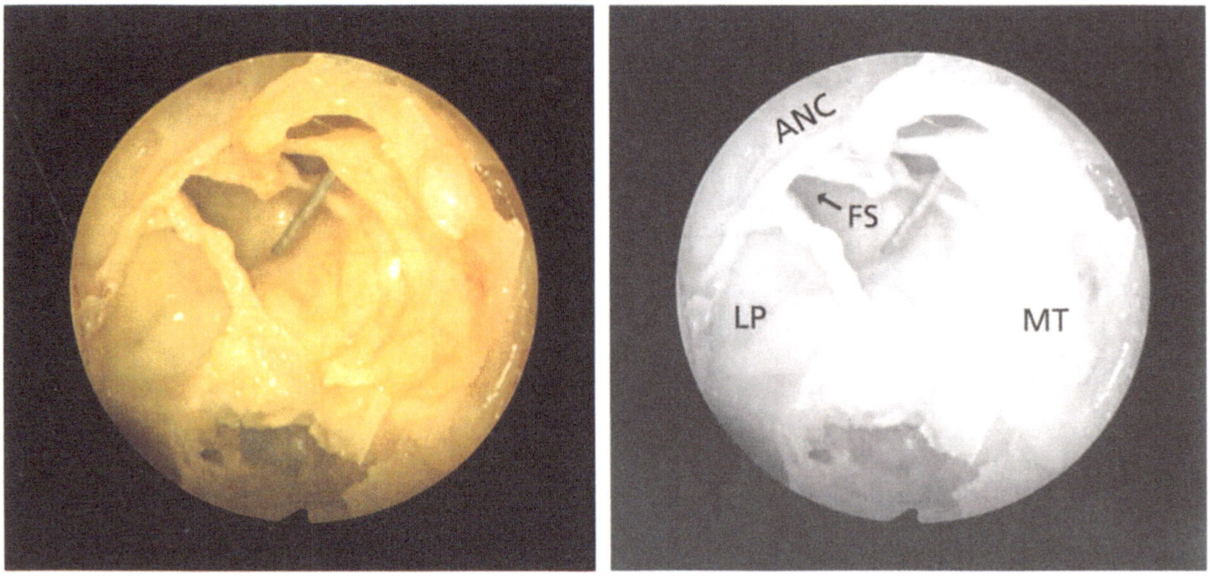

Fig. 17.18. Right side, 70 degree lens, probe from ethmoid frontal.

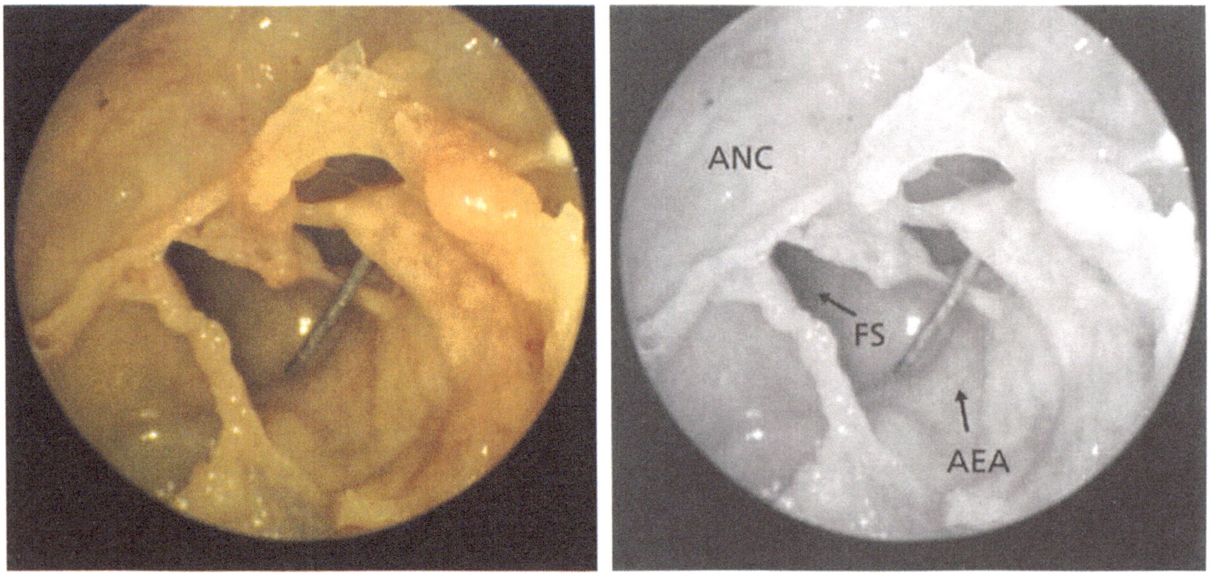

Fig. 17.19. Right side, 70 degree lens, same specimen, close-up.

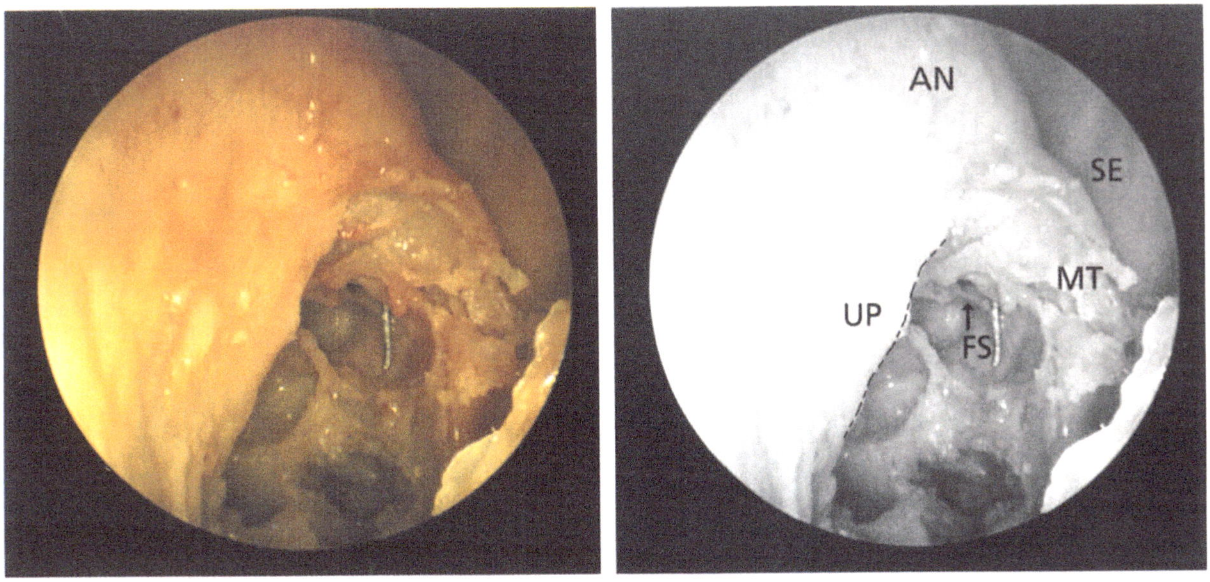

Fig. 17.20. Right side, zero degree lens, probe from frontal sinus.

Glossary

Anterior ethmoid ostium: The opening from an anterior ethmoid cell into the infundibulum.

Dome of ethmoid: The most superior extent of the anterior ethmoid complex.

Exit of infundibulum: The most inferior and posterior portion of the hiatus semilunaris.

Final common pathway: The most inferior portion of the infundibulum between the maxillary sinus ostium and the exit of infundibulum.

Frontal drainage: The space or the potential space behind the posterior and medial wall of the agger nasi cell is the transition space for the frontal sinus.

Hiatus semilunaris posterior: The transition space for the ethmoidal bulla located between the medial wall of the bulla and the middle turbinate.

Infundibulum: The transition space for the maxillary sinus and variably anterior ethmoid and frontal drainage.

Outflow tract: The path taken by drainage from a sinus.

Parson's window: A Setliff modification of Parson's original recommendation for a retrograde approach to the uncinate process. It is a controlled infundibulotomy.

Ribbed vault: A post-surgical finding of skull base.

Sinus lateralis: A variable recess posterior to the ethmoidal bulla.

Skull base: The most superior extent of the bony roof of the ethmoid complex.

Sphenoethmoid recess: A variable space at the most posterior limit of the interface between the middle turbinate and the septum and into which the sphenoid sinus drains.

Tail of the uncinate process: The most posterior and inferior extension of the uncinate process.

Transition space: The space between the sinus ostium and its entry into the nasal cavity.

Bibliography

Becker SP: Applied anatomy of the paranasal sinus with emphasis on endoscopic surgery. *Ann Otol Rhinol Laryngol* 1994; **162**:3-32.

Bolger WE, Butzin CA, Parsons, DS: Paranasal sinus bony anatomic variations and mucosal abnormalities: CT analysis for endoscopic sinus surgery. *Laryngoscope* 1991; **101**:56-64.

Bolger WE, Woodruff WW, Parsons DS: Maxillary sinus hypoplasia classification and description of associated uncinate process hypoplasia. *Otolaryngol Head Neck Surg* 1990; **103**:759-765.

Draf W: *Endoskopie der Nasennebenhöhlen: Technik, typische Befunde, therapeutische Möglichkeiten*. Springer Berlin, Bergmann München, 1983.

Gruenwald L: Deskriptive und topographische Anatomie der Nase und ihrer Nebenhoehlen. In Denker A, Kahler O (eds): *Die Krankheiten der Luftwege und der Mundhoehle.* Springer Berlin, Bergmann München 1925; 1-95.

Hajek M: *Pathologie und Therapie der entzuendlichen Erkrankungen der Nebenhoehlen der Nase*. Deuticke Leipzig, 1909.

Hechl P, Tschabitscher M: Endoscopic Aided Surgery of the Ethmoid Cells. Int. Advanced Course on Functional Endoscopic Sinus Surgery. Vienna, 1992. (Demonstr.)

Kainz J, Stammberger H: Danger areas of the posterior rhinobsis: an endoscopic and anatomical-surgical study. *Acta Otolaryngol* 1992; **112**:852-861.

Kavanaugh KT: Paranasal sinus and anterior skull base: interactive anatomy. *Ann Otol Rhinol Laryngol* 1995; **104**:488-489.

Kuhn FA, Bolger WE, Tisdal RG: The agger nasi cell in frontal recess obstruction: an anatomic, radiologic, and clinical correlation. *Operat Techn Otolaryngol Head Neck Surg* 1991; **2**:226-231.

Lang J: Über die cellulae ethmoidalies posteriores und ihre Beziehung zum Canalis opticus. *HNO* 1988; **36**:49-53.

Levine HL, May M, Rontal MR: Complex anatomy of lateral nasal wall: simplified for the endoscopic sinus surgeon. In Levine HL, May M (eds): *Endoscopic Sinus Surgery*. Thieme Medical Publishers New York, 1993.

Lusk RP, Muntz HR: Endoscopic sinus surgery in children with chronic sinusitis-a pilot study. *Laryngoscope* 1990; **100**:654-658.

Mattox DE: Anatomy of the Ethmoid Sinus. *Otol Clin North Amer* 1985; **18**:3-14.

Messerklinger W: Endoscopic anatomy of the human ethmoturbinals. *Acta Otolaryngol* 1973; **75**:243-248.

Mourel J: Anatomie descellules ethmoidales. *Rev Hebdo de Laryngol Otol Rhinol* 1889; **31**:913-924.

Ohnishi T: Bony defects and dehiscences of the roof of the ethmoid cells. *Rhinology* 1981; **19**:195-202.

Onodi A: The optic nerve and the accessory cavities of the nose. *Ann Otol* 1908; **17**:1-52.

Onodi A: *The Optic Nerve and the Accessory Sinuses of the Nose.* William Wood New York, 1910.

Onodi A: *Accessory Sinuses of the Nose in Children.* William Wood New York, 1911.

Onodi A: *The Relations of the Lacrymal Organs to the Nose and Nasal Accessory Sinuses.* William Wood New York, 1913.

Parsons D, Bolger W, Boyd E: The „Ridge"-a safer entry to the sphenoid sinus during functional endoscopic sinus surgery in children. *Operat Techn Otolaryngol Head Neck Surg* 1994; **5**:43-44.

Parsons DS, Nishioka G: Pediatric sinus surgery. In Kennedy O (ed): *Sinus Disease: Current and Endoscopic Management.* WB Saunders Philadelphia, in press.

Parsons DS, Lockette JS, Martin TW: Pediatric endoscopy: anesthesia and surgical techniques. *Amer J Otolaryngol* 1992; **13**:271-283.

Parsons D, Phillips SE: Functional endoscopic surgery in children: a retrospective analysis of results. *Laryngoscope* 1993; **103**:899-903.

Parsons D, Pransky S: Functional endoscopic sinus surgery in infants and young children. In *Instructional Courses, American Academy of Otolaryngology, Head and Neck Surgery,* Vol 5. Mosby Year Book St. Louis 1992; 159-164.

Parsons DS, Stivers EF: The missed ostium sequence and the surgical approach to revision functional endoscopic sinus surgery. *Otolaryngol Clin. North Amer* 1996; **29**:168-183.

Paturet G: Traité d'Anatomie humaine. Masson.

Rice DH, Schaefer SD: *Endoscopic Paranasal Sinus Surgery*. Raven Press New York, 1988.

Ritter FN: *The Paranasal Sinuses: Anatomy and Surgical Technique,* 2nd ed. C.V. Mosby St. Louis, 1978.

Setliff RC: Minimally invasive sinus surgery: the rationale and technique. *Otolaryngol Clin. North Amer* 1996; **29**(1):115-129.

Setliff RC, Parsons DS: The "Hummer": new instrumentation for functional endoscopic sinus surgery. *Am J Rhinol* 1994; **8**:275-278.

Setliff RC, Parsons DS, Chambers CC: Special considerations in FESS. *Oper Techn Otolaryngol Head Neck Surg* 1994; **5**:40-42.

Stammberger H: Endoscopic endonasal surgery: concepts in treatment of recurring rhinosinusitis. Part I: Anatomic and pathophysiologic considerations and Part II: Surgical technique. *Otolaryngol Head Neck Surg* 1986; **94**:143-146, 147-156.

Stammberger H: *Orientation Important during Functional Endoscopic Sinus Surgery.* Karl-Storz Endoscopy-America, 1987.

Stammberger H: *Functional Endoscopic Sinus Surgery*. Dekker Philadelphia, 1991.

Stammberger H, Kennedy DW (eds): Paranasal sinuses: anatomic terminology and nomenclature. *Ann Otol Rhinol Laryngol* 1995; **104** (Suppl 167):7-16.

Stammberger H, Posawetz W: Functional endoscopic sinus surgery: concept, indications, and results of the Messerklinger technique. *Eur Arch Otorhinol* 1990; **247**:63-76.

Stankiewicz JA: The endoscopic approach to the sphenoid sinus. *Laryngoscope* 1989; **99**:218-221.

Stankiewicz JA: The endoscopic repair of choanal atresia. *Otolaryngol. Head Neck Surg* 1990; **103**:931-937.

Terrier F, Terrier G, Rufenacht D, Friedrich JP: Anatomy of the ethmoid bone region: topographical, radiological and endoscopic landmarks. *Ther Umsch* 1987; **44**:75-85.

Van Alyea OE: *Nasal Sinuses: An Anatomic and Clinical Consideration*. Williams & Wilkins Baltimore, 1942.

Van Alyea OE: Ethmoid labyrinth. *Arch Otol* 1939; **29**:881-902.

Vleming M, Middelweerd RJ, de Vries N: Complications of endoscopic sinus surgery. *Arch Otolaryngol Head Neck Surg* 1992; **118**:617-623.

Wigand ME: *Endoscopic Surgery of the Paranasal Sinuses and Anterior Skull Base*. Thieme Stuttgart/New York, 1990.

Wolf G, Anderhuber W, Kuhn F: Development of the paranasal sinuses in children: implications for paranasal sinus surgery. *Ann Otol Rhinol Laryngol* 1993; **102**:705-711.

Zinreich SJ, Mattox DE, Kennedy DW, Chisholm ML, Diffley DM, Rosenbaum AE: Concha bullosa: CT evaluation. *J Comput Assist Tomogr* 1988; **12**:778-784.

Zuckerkandl E: Die Siebbeinmuscheln des Menschen. *Anat Anz* 1892; **7**:13-25.

Index

* Page numbers for figures are given in parentheses.